Nietzsche's Posthumanism

CARY WOLFE, SERIES EDITOR

68 *Nietzsche's Posthumanism*
Edgar Landgraf

67 *Subsurface*
Karen Pinkus

66 *Making Sense in Common: A Reading of Whitehead in Times of Collapse*
Isabelle Stengers

65 *Our Grateful Dead: Stories of Those Left Behind*
Vinciane Despret

64 *Prosthesis*
David Wills

63 *Molecular Capture: The Animation of Biology*
Adam Nocek

62 *Clang*
Jacques Derrida

61 *Radioactive Ghosts*
Gabriele Schwab

60 *Gaian Systems: Lynn Margulis, Neocybernetics, and the End of the Anthropocene*
Bruce Clarke

59 *The Probiotic Planet: Using Life to Manage Life*
Jamie Lorimer

58 *Individuation in Light of Notions of Form and Information*
Volume II. *Supplemental Texts*
Gilbert Simondon

57 *Individuation in Light of Notions of Form and Information*
Gilbert Simondon

(continued on p. 261)

Nietzsche's Posthumanism

Edgar Landgraf

posthumanities 68

University of Minnesota Press
Minneapolis
London

Portions of chapter 2 are adapted from "The Physiology of Observation in Nietzsche and Luhmann," *Monatshefte* 105, no. 3 (2013): 472–88, copyright 2013 by the Board of Regents of the University of Wisconsin System, reprinted courtesy of the University of Wisconsin Press. Portions of chapters 3 and 4 are adapted from "Nietzsche's Entomology: Insect Sociality and the Concept of the Will," *Nietzsche-Studien* 50, no. 1 (2021): 275–99, https://doi.org/10.1515/nietzstu-2021-0011.

Copyright 2023 by the Regents of the University of Minnesota

All rights reserved. No part of this publication may be reproduced, stored in a retrieval system, or transmitted, in any form or by any means, electronic, mechanical, photocopying, recording, or otherwise, without the prior written permission of the publisher.

Published by the University of Minnesota Press
111 Third Avenue South, Suite 290
Minneapolis, MN 55401–2520
http://www.upress.umn.edu

ISBN 978-1-5179-1532-2 (hc)
ISBN 978-1-5179-1533-9 (pb)

A Cataloging-in-Publication record for this book is available from the Library of Congress.

The University of Minnesota is an equal-opportunity educator and employer.

Contents

Acknowledgments   *vii*
Abbreviations   *ix*
Introduction   *xi*

1 Posthumanism and Its Nietzsches   *1*
2 Posthumanist Epistemology   *25*
3 Insect Sociality   *59*
4 Instinct, Will, and the Will to Power   *79*
5 Media Technologies of Hominization   *101*
6 Cultivating the Sovereign Individual   *125*
7 The Ethics and Politics of Nietzschean Posthumanism   *151*

Notes   *185*
Bibliography   *223*
Index   *241*

Acknowledgments

Speaking of distributed agency: even with an extensive bibliography, it is impossible to acknowledge appropriately everyone and everything that goes into the writing of a book. Staying with convention, then, I will acknowledge (and not hold responsible for any shortcomings of this book) only the most immediate professional and personal actors.

First, I thank Cary Wolfe, editor of the Posthumanities series, for his strong support of the book proposal that pushed me to bring the project to completion, and for his eager endorsement of the manuscript that ensured its inclusion in the series. The latter was also aided much by Stefan Herbrechter's generous, detailed, and helpful review of the manuscript, for which I am very grateful. I also thank Christian Emden for his continued professional support and, along with him, the anonymous reviewer of *Nietzsche Studien,* for feedback on a condensed version of the insect chapters. Mike Stoffel and the editorial team at the University of Minnesota Press did an amazingly thorough job copy editing the final manuscript for consistency and style, for which I am very grateful.

The project also benefited from the input of many of the students who attended seminars I taught on posthumanism, on Nietzsche, and on philosophies of technology. Without their probing questions and ideas, the book would not be the same. I would also like to acknowledge the importance of the German Studies Association introducing research seminars as an alternative to presentation panels at its annual conferences, and to specifically thank the organizers and participants of the 2017 GSA seminar "*Technologie*: Readings in a Neglected Discipline" in Atlanta and of the 2019 GSA seminar "*Kulturtechnologie*" in Portland, Oregon. These seminars provided the theoretical groundwork for chapter 5 and helped frame the argument developed in chapter 6.

Among the organizers of these seminars, I want to extend special thank-yous to Leif Weatherby and Gabriel Trop, who also conspired on coediting the volume on *Posthumanism in the Age of Humanism* (Bloomsbury, 2019), which helped me refine my understanding of the significance of nineteenth-century science and philosophy for posthumanism. So happy to have you as colleagues who keep pushing the bounds of German studies and, more important, to have you as friends!

We share as a dear friend and coconspirator Jocelyn Holland, who has been at the center of a network of professional and social activities that have been enlivening my intellectual life since graduate school. *Nietzsche's Posthumanism* would not have come together without your friendship, continued encouragement, and extensive feedback on early (and last-minute) drafts of each of its chapters.

The book would also not be what it is without another dear friend and accomplice, Clayton Rosati. Thank you, Clayton, for your careful readings of chapters 5 and 7 and for your many suggestions. More important, thank you for enduring hours and hours of my gradually constructing thoughts while speaking, for your countless probing questions and your unsolicited but much appreciated advice, and for continually holding my feet to Marx's fire.

Finally, I could have not written this book without the extensive support of friends and family. In particular, I thank Clemens Berger, Todd Herzog, and Rob Wallace, as well as Phil Stahel and my brother Wilbert for their friendship, perspective, and humor. Last but not least, the book owes its existence also to the patience, support, and love of my sons Allan and Alex, who make me proud beyond words, and of my wife, Angela, the love of my life.

Abbreviations

Translations of Nietzsche's work are cited using these abbreviations, followed by the section or aphorism number, unless the number is preceded by a comma, indicating a page number. For the German original, I consulted the *Kritische Studienausgabe* (KSA), edited by Giorgio Colli and Mazzino Montinari, 15 volumes. References to Nietzsche's unpublished notebooks not contained in the KSA follow the *Kritische Gesamtausgabe* (KGW), edited by Giorgio Colli and Mazzino Montinari. Citations reference part, volume (within part), and page number (e.g., KGW, 2/2:188 = KGW, part 2, volume 2, at page 188).

All translations of works not readily available in English are mine.

A   *The Antichrist*
BGE   *Beyond Good and Evil*
BT   *The Birth of Tragedy*
D   *Daybreak/Dawn*
EH   *Ecce Homo*
G   *On the Genealogy of Morals*
GS   *The Gay Science*
HH   *Human, All Too Human*
HL   *On the Uses and Disadvantages of History for Life* (the second UM)
KGW   *Kritische Gesamtausgabe*
KSA   *Kritische Studienausgabe*
KSB   *Kritische Studienausgabe Briefe*
TI   *Twilight of the Idols*
TL   *On Truth and Lies in a Nonmoral Sense*

ABBREVIATIONS

UM *Untimely Meditations*
WS *The Wanderer and His Shadow* (part 2 of volume 2 of HH)
Z *Thus Spoke Zarathustra* (references list part number and chapter titles)

Introduction

PUTTING A SINGLE AUTHOR at the center of a posthumanist inquiry must seem to be caught in a performative contradiction. How can you challenge humanism's anthropocentrism by centering a book around the writings of an individual human being? How can you challenge the biological, epistemological, and social privileging of humans by reaffirming and extending the privileged position of a single author, an author, no less, who is known for concepts such as the will to power and the *Übermensch* that appear to express hyper forms of human agency and exceptionality? The quandary presents but a narrower version of a contradiction any alert student of posthumanism will notice sooner or later. Are not humans those who articulate the posthuman, and who do so—for whom else?!—but for humans? The predicament applies more generally to posthumanism's relationship to humanism. Kate Soper warns that, in trying to run away from humanism, in trying to overcome its dichotomies, posthumanists only reinforce these dichotomies: "In the last analysis, the rhetorical address of the posthuman is always a call that makes sense only as an appeal to those already within the human community."[1]

There are two ways to resolve this apparent contradiction. On the side of content, we may consider that the point of posthumanism is not to run away from humanism, but to redefine "the human," and by extension, as Jane Bennett puts it, to rearticulate "operative notions of matter, life, self, self-interest, will, and agency."[2] It is easy to see how Nietzsche's writings contribute to such a project, as they pressure the very concepts Bennett lists as targets of posthumanist inquiry. A second way to escape posthumanism's performative self-contradiction requires a more far-reaching conceptual intervention. It entails a reimagining of knowledge, meaning, and even human thought in ways that are no longer anchored in a human subject or mind, or in self-consciousness. Nietzsche's writings also advance this perspective, be it through their deconstruction of notions of self, the subject, or the "soul," their focus on the body, their exploration of the limits of language, or through their aphoristic, fragmentary, and poetic styles that, at a minimum, redistribute auctorial control from the author to the reader, soliciting from the latter a good amount of interpretive creativity.[3]

This book pursues both of these lines of argument. It examines how Nietzsche challenges the centrality and exceptionality of the human subject and the human species while offering new modes of thinking that break with central humanist conceits. It will further lessen the inherent anthropocentrism of a book-length study on a single author by reading Nietzsche's writings not in a vacuum, not as anchored and legitimized by what we might divine were the real intentions of the mind that put them on paper, but as part of a larger tapestry formed by intersecting philosophical and scientific discourses. The latter, which have been the focus of an extensive body of research in recent decades, must be of particular interest for a study of Nietzsche's posthumanism. For, just as major tenets of Nietzsche's thought formed in conversation with findings from the natural and life sciences,[4] the confluence of science and philosophy is also a central feature of contemporary posthumanism. In fact, Nietzsche's studies of the sciences of his day focus precisely on the kind of research that is critical to contemporary posthumanism: materialism, physiology, neurophysiology, ethology, and even entomology, which already in the nineteenth century had come to reflect on the strange dynamics of swarms that it also applies to human sociality.

Along with its interest in the natural sciences, posthumanism is defined by its focus on technology. Posthumanists study technology not just for its social, cultural, and political effects, but are interested in determining how technology displaces traditional borders between the organic and the inorganic, between nature and culture, changing the cognitive capabilities of humans along with their physiology. In this regard, the cyborg has become the unofficial symbol of posthumanism. The cyborg raises the question of what it means to be human in a world so thoroughly shaped by technology. As with the natural sciences, philosophical reflections on the effects of technology, too, date back to the nineteenth century. In fact,

Nietzsche lived in a time that witnessed unprecedented scientific discoveries and technological advances when, as Gregory Moore details with regard to the period between 1820 and 1900, "an overwhelmingly rapid succession of innovations transformed the world and the way people understood it: in engineering, the development of mechanized industry, the railways, telegraphy, synthetic dyes, the telephone, gas and electric lighting; in physics, the articulation of thermodynamics and the rise of statistical mechanics; in the life sciences, the discovery of cell, evolutionary theory, and experimental physiology."[5] While the question of technology has received less attention in the secondary literature on Nietzsche, his reflections on technology and those of his time focus on some of the same questions contemporary posthumanists ask: How do technologies, particularly communication-media technologies, affect humans socially, culturally, but also psychologically and physiologically? How do they help constitute and alter humans' sensibilities, their self-image, and their behavior?

That a closer examination of developments in science and technology in the nineteenth century will help uncover continuities between Nietzsche's thought and contemporary posthumanism should not be surprising. The nineteenth century laid the foundations for the experimental sciences, from chemistry and physics to biology, ethology, and sociology, and its technological inventions continue to be used today. Inasmuch as contemporary science still stands on the shoulders of many of the nineteenth-century thinkers Nietzsche read, so does the posthumanist literature that draws on extensions of these scientific traditions. A central concern for scientists then and now, one that unavoidably raises philosophical points of contention, is their attempt to avoid the intrusion of metaphysical presuppositions into their explanatory models. Nietzsche's philosophy participates specifically in science's drive to rid itself of hidden conceptions of the soul (which Nietzsche still detects in the concept of the atom, for example) and from teleological modes of thinking that attribute causal force (*causa efficiens*) to the outcome of complex processes, and to purge itself of persistent dualisms such as those between mind and body and between mind and matter. Nietzsche reflects as much on the metaphysical propositions that continue to guide scientific inquiry as on the challenges the scientific developments of his age impose on philosophy. Then and now, the dialogue between science and philosophy forces a fundamental rethinking of the status of the human, of human cognition, and of the relationship of human beings to each other, as well as to animals, the environment, and the realm of the inorganic.

By focusing on the scientific discourses that inform Nietzsche's writings and on his reflections of technology, this book hopes to offer a more systematic assessment of the relevance of Nietzsche's thought for posthumanism. Such an assessment is valuable on two accounts. Firstly, it promises to add an important chapter to the genealogy of posthumanism while also shedding light on some of posthumanism's own internal tensions, such as those about the persistence of naturalist presuppositions, or of humanist conceits about human agency and morality. Secondly, putting Nietzsche in dialogue with posthumanist theories opens up new vistas onto Nietzsche's thought, giving new emphasis to aspects of his writings that have been less well known. These include the neurophysiological basis of Nietzsche's epistemology, his concerns with insects and the emergent social properties they exhibit, and his reflections on the hominization effects of technology. The ethos of critical posthumanism also offers a new perspective on some of the key moral and political contentions of Nietzsche's writings.

By putting Nietzsche in dialogue with posthumanism and vice versa, this book will necessarily be selective rather than comprehensive in its reading of Nietzsche's philosophy, as well as in its discussion of the highly diverse literature on posthumanism. It will focus on topics where the continuities and points of contentions between Nietzsche and posthumanism are most promising and ignore other areas where there is little or only sporadic continuity between the two. Despite laying no claim on comprehensiveness, the chapters of this book hope to make unique contributions to the existing scholarship while also presenting uniquely framed introductions to some of the main tenets of Nietzsche's thought and of contemporary posthumanism. Chapter 1, "Posthumanism and Its Nietzsches," will offer a broader reflection on posthumanism as a shifting and by no means unified field of inquiry—a diversity that is reflected in the variety of different Nietzsches posthumanists have claimed for themselves. The focus of the overview is an underlying onto-epistemological divide that, simply put, separates strands of posthumanism that continue to essentialize human nature, on the one hand, from those that extend the lessons of poststructuralism and avoid such essentializations, on the other. The division appears most clearly in differing attitudes toward the Enlightenment. *Critical* posthumanists such as N. Katherine Hayles, Stefan Herbrechter, or Cary Wolfe use the Enlightenment as a measuring stick to differentiate theirs from strands of posthumanism that extend rather than refute major tenets of Enlightenment thought.

To avoid the common conflation of humanism with Enlightenment thought, however, chapter 1 heeds Michel Foucault's suggestion that "just as we must free ourselves from the intellectual blackmail of being for or against the Enlightenment we must escape from the historical and moral confusionism that mixes the theme of humanism with the question of the Enlightenment."[6] Foucault's point is that the critical impetus of the Enlightenment is in principle opposed to the normative tendencies inherent to the wide variety of past and present humanisms (from the humanism of Christianity and Marxism to the humanist contentions of national socialism and Stalinism). To profile more clearly the humanism that is the target of Nietzsche's critique and that is distinct from the critical consciousness of the Enlightenment, chapter 1 turns to Martin Heidegger's *Letter on Humanism,* a text that offers a more expansive definition of humanism that aligns with Nietzsche's broad critique of the trajectory of Western thought since Socrates. Heidegger's understanding of humanism is also interesting for contemporary posthumanist concerns, as it defends explicitly a posthumanist ethos that recognizes how "the humanistic interpretations of man as *animal rationale,* as 'person,' as spiritual-ensouled-bodily being, . . . do not realize the proper dignity of man."[7] While Heidegger himself, sadly, falls short of the latter, his analysis nevertheless helps clarify how a critical posthumanism need not reject ideas of progress, nor does it have to jettison the idea that its ethos could and should aim toward a betterment of the human lot in terms of a reduction of violence and discrimination. Rather, its aim is to better understand and avoid the reasoning and values that have prevented humanism to live up to many of its most cherished ideals.

Chapter 2, "Posthumanist Epistemology," revisits Nietzsche's epistemology in relation to the life and neurophysiological sciences of his time, taking on both recent ontological contentions by new materialists and more familiar poststructuralist interpretations of Nietzsche. In the light of the scientific research Nietzsche consulted, the chapter argues that Nietzsche extends a neo-Kantian epistemology that should not be viewed as "correlationist," as some of the new materialists put it, but rather as one that follows a nineteenth-century neurophysiological tradition that developed early models of embodied or *enactive* cognition along the lines described in the twentieth century by Humberto Maturana and Francisco Varela.[8] Particularly, chapter 2 shows how Nietzsche's epistemological contentions build on Johannes Müller's principle of specific nerve energies, which states, in a nutshell, that the output our senses produce does not reflect anything qualitative or "essential" about the input that the sensory apparatus processes, but reflects merely the specific capabilities of the human physiology. The principle of specific nerve energies forms the basis of a scientific tradition that understands consciousness and cognition as emergent phenomena tied to the dynamics of self-referencing neural networks and reciprocally interacting systems. Müller, and by extension Nietzsche, thus contribute to an episteme that offers an alternative to reductionist and materialist approaches that came to dominate the life sciences in the nineteenth century. Revisiting the neurophysiological literature Nietzsche consulted makes apparent how Nietzsche's epistemology remains relevant for contemporary debates on embodiment and even offers an alternative epistemological model that is neither ignoring Kant for a supposedly unmediated ontology or a new (speculative) realism nor extending the anthropocentrism of what some of the new materialists chastise as "correlationism."

Chapter 2 concludes with a reflection on how a cybernetically informed epistemology is able to break the theoretical and political logjam between critical posthumanists and some of the new materialists who oppose epistemology in the vein of Kant altogether and who suggest that it is on the "terrain of ontology that many of the urgent ecological battles need to be fought."[9] Against this position, chapter 2 argues that the organic or operational constructivism that connects Nietzsche to Müller and twentieth-century neocybernetic epistemologies does not lead to a denial of the reality of scientific findings, but only of their grounding in an absolute, unchangeable outside. The interpretive openness of Nietzsche's writing thus reflects a constructivist viewpoint that rejects ontological or materialist reductionisms, advocating instead for a continued differentiation of linguistic, scientific, and technological processes that can adapt to and engage the urgent ecological battles humanity faces.

Aspects of a cybernetic tradition of posthumanist thought are also at the heart of chapter 3, "Insect Sociality," and chapter 4, "Instinct, Will, and the Will to Power," which examine Nietzsche's references to insects and the entomological research that informs them. Insects appear quite regularly in the literature on posthumanism, though so far only Jussi Parikka, drawing on the works of Gilles Deleuze and Félix Guattari, has offered a detailed analysis of the confluence between entomology and posthumanism. Already in the nineteenth century, entomologists and ethologists—including the French thinker Alfred Espinas, whose book *On Animal Societies* from 1877 Nietzsche marked heavily—observed in insect hives behavior that challenged the privileging of higher (human) faculties. The amazing accomplishments of insect societies point toward what Hayles calls a "cognitive nonconscious"[10] and offer models for what today we call emergent phenomena. What insect hives make apparent in the nineteenth century are organizations where interacting elements constitute a whole that exhibits properties that cannot be attributed to the elements or actors independent of their interactions.

While ants and other social insects had caught the attention of philosophers as far back as Aristotle, in the nineteenth century they were no longer of interest merely as social or political analogues. In the eyes of entomologists, they offered insights into the basic nature of instincts and of human sociality as well as that of animals. Both questions go to the heart of Nietzsche's philosophy. Insect behavior, broadly put, demonstrates the efficacy of instinctual and nonconscious motivators in the absence of higher cognitive faculties, an anthropological presupposition that Nietzsche sustains throughout his writings. Furthermore, because of the discrepancy between the achievements of the collective and the limited cognitive and communicative abilities of the individual members of the collective, insects raise fundamental questions about the relationship between the wants, desires, and reasoning of individual members of a species and about the functionality of the whole that forms from their interactions. Nietzsche is, of course, also no stranger to this second matter. The focus on insects thus offers another opportunity to read Nietzsche, as Vanessa Lemm suggests, through the lens of an "affirmative biopolitics" that "sees in the continuity between human and animal life a source of resistance to the project of dominating and controlling life-processes."[11] Furthermore, examining Nietzsche's entomology allows us to add a posthumanist perspective to the more familiar poststructuralist readings of Nietzsche, and conversely to add historical and philosophical context to contemporary posthumanist concerns with emergent properties and environmental embeddedness.

Against the backdrop of Nietzsche's understanding of human sociality based on that of insects as elaborated in chapter 3, chapter 4 revisits his conceptions of instinct, the will, and the will to power. At the center of the chapter stands an unpublished note from 1883 that compares the physical interactions of ants with those of colliding wills, arguing that it is from such rudimentary exchanges that we must derive the most basic modes of understanding and the formation of basic social aggregates. The note shows how Nietzsche's deconstruction of the will as an umbrella concept and his will to power are attempts to model the emergence of complex edifices on simple operations under which physical, psychological, and social phenomena must be thought to arise. Nietzsche's argument builds on the psychophysiological writings of the physiologist Wilhelm Wundt. Nietzsche followed Wundt's writings over decades. They elucidate how Nietzsche's will to power reflects physiological discourses of his time that aim to rid the sciences of dualistic and metaphysical modes of thinking. Nietzsche takes from Wundt a conception of "will" that is interesting for posthumanists because it cuts across species and even across the divide between organic and inorganic. Nietzsche's will to power, the chapter maintains, thus anticipates the kind of material vitalism Deleuze and Guattari inspired in Bennett's seminal book *Vibrant Matter*.

Chapter 4 concludes with a reflection on the social and political relevance of what Nietzsche identifies as the "disgregation of the will" in modern society. It argues that Nietzsche recognizes how affirmations of the will do not contest, but only help solidify the formation of herds. This is not merely an academic problem if we consider the political exploits of the "cult of the will" (as Michael Cowan titles his study) in the first half of the twentieth century. The entomological research of his time allows Nietzsche to think of social orders as self-organizing aggregates—in a famous aphorism on the "impossibility" of the working class, he likens modern Europe to a beehive—rather than along the lines of a dichotomy between individuals and societal forces (anticipating Bruno Latour's critique of a central conceit of modern sociology). This understanding of societal orders challenges popular interpretations of Nietzsche that reduce his social theories to dichotomies between the strong and the weak, between the sovereign individual and the herd, and so on. On closer inspection, such dichotomies collapse, given that, to speak with Jacques Derrida, "Nietzsche himself cannot prevent the most puny weakness being at the same time the most vigorous strength."[12]

The cyborg has become the unofficial symbol of posthumanism, expressing the extensive interest of many of its strands in how technology alters what it means to be human and how it might affect the future of the species. Against this backdrop, chapter 5, "Media Technologies of Hominization," and chapter 6, "Cultivating the Sovereign Individual," examine Nietzsche's reflections on technology, a topic that has received relatively little attention in the secondary literature. This neglect is best explained by the narrow definition of technology dominant today. Chapter 5 therefore first sets out to trace changes in the notion of technology in the nineteenth century, including in the writings of Karl Marx and Ernst Kapp, the two most important philosophers of technology in Nietzsche's time. Both thinkers view technology as contributing in more fundamental ways to the process of "hominization" (Bernard Stiegler) and human cultivation than contemporary assumptions about gadgets, scientific contraptions, and even biotechnologies would lead us to believe. From a critical posthumanist perspective that acknowledges the constitutive role of technologies on human evolution, chapter 5 revisits Nietzsche's reflection on the culturally problematic effects of communication-media technologies, specifically on the new "barbarism" allegedly fostered by the mass media of his time, the printing press, and by mass education. Because Nietzsche does not reduce technology to gadgets, add-ons, or prostheses that are understood as mere extensions of the human, the chapter argues, he is able to appreciate how intricately involved technologies are in constituting human subjectivity and agency. Revisiting the great divide elaborated in chapter 1, chapter 5 concludes with a reflection of Donna Haraway's cyborg figure, which extends the critical lessons of Nietzsche's posthumanism as her analysis exposes and challenges the normative thrust of the Western humanist tradition, including, but not limited to, its definitions of gender.

A posthumanist focus on technology as outlined in chapter 5 makes it possible to bring together two seemingly disparate aspects of Nietzsche's writing: on the one hand, his reflections on the culturally ruinous effects of communication-media technologies that anticipate ideas that are fully spelled out only by the fathers of modern media theory, E. A. Havelock, W. J. Ong, and Marshall McLuhan, and on the other, his most extensive elaboration of the cultivation processes that produced us "modern" humans as laid out in *On the Genealogy of Morals*. In that text, Nietzsche treats language itself, the structures and hierarchies it introduces, as a technology of hominization and cultivation. Furthermore, Nietzsche identifies a basic psycho-economic calculus, a "most primitive kind of cunning," at the base of what "was also, presumably, the first appearance of human pride, man's sense of superiority over other animals" (GM II: 8). This calculus remains central in modern civic societies and for the "sovereign individual," the figure Nietzsche identifies as the ultimate product of the humanist cultivation processes (GM II: 1). Chapter 6 argues that, by focusing on linguistic operations and the evolution of particular interpretive schemas, Nietzsche puts forward a more radical notion of embeddedness of humans and human agency, of self and subjectivity, than can be found in the writings of many contemporary posthumanists. Moreover, Nietzsche's genealogical account of humanism's cultivation process reveals how existing moral codes and ideals extend rather than overcome modes of "human, all too human" thinking and valuing that promote rather than overcome violence. This also applies to institutions of justice that higher powers employ to combat resentment and to control the violence at the heart of the rudimentary economic calculus from which humankind emerged. By virtue of their target, they remain tied to the compensatory logic they combat. Chapter 6 concludes by engaging tensions between naturalist readings of Nietzsche and "assemblage posthumanism" as raised by Lemm's 2020 book *Homo Natura*. An expanded notion of technology more appropriate for the nineteenth century can break the impasse between Lemm's dismissal of technology in her assessment of Nietzsche's posthumanism and Rosi Braidotti's paradoxical affirmation of technology for the purpose of a renaturalization of posthuman subject formations.

Finally, chapter 7, "The Ethics and Politics of Nietzschean Posthumanism," returns to Nietzsche's ethical positions and some of his problematic political assertions. The first part of the chapter engages recent warnings from the left about the "dangerousness" of Nietzsche's mind, such as by Malcom Bull and Ronald Beiner, based on a renewed interest in Nietzsche on the part of far right and alt-right movements. Rather than try to rescue Nietzsche from his fascist appropriations or his liberal critics, however, the chapter examines in what respect Nietzsche's posthumanism offers an outside perspective on these political debates. Roberto Esposito has gone perhaps furthest in profiling the problematic core of Nietzsche's political contentions while also recognizing the deconstructive force of Nietzsche's argument. In his seminal book *Bios*, Esposito credits Nietzsche for "understanding that in the centuries to come the political terrain of comparison and battle will be the one relative to redefining the human species in a scenario of progressive displacement of its borders with respect to what is not human, which is to say, on the one hand to the animal and on the other to the inorganic."[13]

The deconstructive force of Nietzsche's argument is similarly on display in his critique of democracy, a critique that chapter 7 argues should be of interest to critical posthumanists, as it offers insights into the apparent failure of the humanist tradition to live up to many of its most cherished ideals, including the creation of a society that would promote more successfully equality, individual freedom, and human dignity. The question gains in significance for an inquiry into Nietzsche's posthumanism because his rejection of Judeo-Christian morals and his well-known affirmations of power and structures of domination seem to contradict a contemporary posthumanist ethos that puts the "emphasis on the respectful co-existence of different perspectives, individuals, groups, and systems,"[14] or one that is viewed as expressing "a grounded form of accountability, based on a sense of collectivity and relationality, which results in a renewed claim to community and belonging by singular subjects."[15] To address the tension between the agonistic structure of Nietzsche's thought and contemporary posthumanism's communal ambitions, the chapter returns to the psycho-economic calculus Nietzsche places at the heart of the humanist cultivation process. A posthumanist perspective offers an alternative interpretation of the frequently used argument that it is resentment that drives much of today's populist (and antipluralist) political support. Not only is such an interpretation overly simplistic, but it also conflates political positions with psychological states in ways that ignore Nietzsche's much more intricate theory of sociality that the present book traces in chapters 3 and 4. The problem is that the psycho-economic calculus that continues to drive our modern legal and political institutions (left and right) makes it difficult to address some of today's most pressing political challenges (inequality, migration, climate change) because, in a nutshell, these problems cannot be solved by means of political, judicial, or even economic compensation.

Contemporary posthumanist articulations of community that are based on notions of a shared debt (Lemm) or ideals of togetherness (Braidotti) encounter a different problem. Such ethical stances continue to put too much emphasis on the reasoning and moral rectitude of individuals. They are also ill-equipped to deal with opposition. Because of these problems, chapter 7 returns to Nietzsche's earlier consideration on sociality and suggests replacing the venerable term "community" with the less revered "swarm." The advantage of the latter concept is that it can better account for the contentious dynamics of social interactions, and for the simultaneous dependence and independence of individuals vis-à-vis the social and technological systems in which they find themselves embedded. The chapter concludes with reflections on how a Nietzschean critique of morality might contribute to a posthumanist ethos that avoids compensatory valuation regiments and acknowledges the multiple embeddedness of post/humans—namely, by differentiating ethical standards beyond good and evil that invite a diversity of viewpoints and help increase the flexibility of the emerging social and material networks they subtend.

Finally, a word on the chosen methodology of this book. Unlike the theory-oriented (or theory-averse) schools of thought that dominated academia over the last fifty years, posthumanism has neither developed nor agreed to adopt a particular theoretical framework for itself. This lack of commitment allows this book to juxtapose rather different strands of posthumanism with different aspects of Nietzsche's thought, some that align, others it deems as misaligned with it. The use of different theoretical paradigms, however, does not mean that this book adopts a neutral stance toward its chosen methodologies. Rather, it is committed to critical, methodological, and mediating strands of posthumanism, strands that, as chapter 1 discusses in more detail, do not essentialize notions of human nature in a dystopic or idealized manner. Furthermore, while some strands of posthumanism distance themselves from the insights of poststructuralism, this book will build on them, even when pursuing new-materialist lines of argument. Put more bluntly, this book subscribes to an ethos that critical posthumanism shares with poststructuralist thinkers such as Derrida or Deleuze and Guattari and with strands of methodological posthumanism, including neocybernetics, Latour's actor-network theory (ANT), and systems theories like that of Niklas Luhmann. These schools of thought aim to avoid any anchoring of their findings in "human nature" for epistemological and ethical reasons, as they recognize the alterity and limits of different modes of cognition and sense-making, including about what defines human nature and the potentially discriminatory effects of such assertions. The wager is that these strands of posthumanism are most compatible with Nietzsche's thought and also most apt to offering new vistas on his writings. What this book hopes to substantiate, then, is the pertinence of these critical posthumanist modes of inquiry for our understanding of Nietzsche's work, and of Nietzsche's work for the understanding and continued development of these critical modes of posthumanist inquiry.

1
Posthumanism and Its Nietzsches

> Hitherto the *lie* of the ideal has been the curse of reality.
> —Friedrich Nietzsche, *Ecce Homo*

IN *POSTHUMANISM,* Stefan Herbrechter suggests that it "would indeed be difficult to overestimate the Nietzschean influence on posthumanism" (21). Certainly, posthumanists of different stripes like to reference and quote Nietzsche, and some even claim him as one of their own. Yet, his reception is as varied as the motley group of schools of thought associated with posthumanism or one its variants. Herbrechter himself identifies two Nietzsches relevant for posthumanism: the "critical Nietzsche," whose ethos "has been informing so-called 'French-Theory,' or poststructuralism and deconstruction," and Nietzsche the "'prophetic' vitalist," who is "craving the coming of the overman," the "metaphysician who seems to blindly follow an absolute 'will'" (22). The "critical Nietzsche" is "intent on breaking up traditional and venerable but ossified knowledge" and "no longer accepts any final forms of truth" (32), while Nietzsche the "prophetic vitalist," with his concept of the overhuman, drives much of Nietzsche's popular legacy and seems to foreshadow contemporary fantasies about an enhanced posthuman or transhuman, a being who will have escaped the limits of the "human, all too human."

More "Nietzsches" relevant for posthumanism could easily be identified within and beyond the two Herbrechter names: the Nietzsche who, to quote the first statute of Robert Pepperell's "Posthuman Manifesto," recognizes "that humans are no longer the most important thing in the universe";[1] the Nietzsche who promotes the "radical decentering of the traditional sovereign, coherent and autonomous human";[2] the

Nietzsche who challenges dualistic modes of thinking, including the mind–body dualism; the Nietzsche whose epistemology avoids the pitfalls of what Quentin Meillassoux calls the "Kantian catastrophe";[3] the Nietzsche whose philosophical anthropology rejects "anthropocentrism, anthropomorphism and species hierarchy based on an idea of a continuum between nature and culture";[4] the Nietzsche who contests humanist strategies of universalization, including the aim to define universally applicable values, laws, and goals; the Nietzsche who recognizes the technological and social embeddedness of humans and who tries to grasp the complex social dynamics that determine human desires, drives, and wills; the Nietzsche who inspired notions of "nomadic becoming";[5] the Nietzsche who searches for ethical values beyond the Judeo-Christian moral codes that dominate the humanist tradition; or the Nietzsche who understands that, as Roberto Esposito puts it, "in the centuries to come the political terrain of comparison and battle will be the one relative to redefining the human species in a scenario of progressive displacement of its borders with respect to what is not human, which is to say, on the one hand to the animal and on the other to the inorganic."[6]

Despite such apparent continuities between Nietzsche and posthumanism, there is disagreement about Nietzsche's historical relevance for posthumanism. While Herbrechter puts Nietzsche at the beginning of "A Genealogy of Posthumanism,"[7] Nick Bostrom, the leading proponent of transhumanism (now "humanity+"), in his "History of Transhumanist Thought," sees only "surface-level similarities" between Nietzsche and transhumanism (4).[8] Indeed, even a superficial look will just as quickly reveal tensions and incongruences between Nietzsche's philosophy and some of the main concerns and outlooks associated with posthumanism. For example, while Nietzsche's critique of humanist values and of humanism's inherent anthropocentrism is pathbreaking, still his notion of the will to power—which, according to Robert Pippin, almost amounts to a "psychologizing of being itself"[9]—appears, at least on the surface, to rescue an anthropocentric viewpoint, and thus might seem like a mere ploy, as German sociologist Niklas Luhmann puts it, "to retain the ability to attribute behavior to individuals."[10] In a similar vein, while Nietzsche challenges us to rethink notions of the self, of subjectivity, of agency in ways that are shared by *critical* posthumanists, how Nietzsche's concept of the overhuman relates to transhumanist ideas about technologically and biotechnologically altered transhumans or posthumans remains the subject of contentious debates.[11]

Associations of Nietzsche with posthumanism get even thornier if we think of parallels that address ethical questions. Nietzsche's philosophical anthropology, for example, challenges essentializations of the difference between humans and animals,[12] yet it is hard to fathom how the defense of animal rights so central to critical posthumanists like Cary Wolfe might be part of Nietzsche's political horizon. The contemporary relevance of Nietzsche's ethics in any case is difficult to assess. Nietzsche's unveiling of power assertions behind even the loftiest of ethical ideals is in line with the political ethos of critical posthumanists such as Herbrechter or Wolfe, yet Nietzsche's often bellicose rhetoric is harder to align with posthumanism's support for a "situated pluralism and diversity"[13] and its desire to extend values of empathy, compassion, and respectfulness within and beyond the human species. It will in any case require an in-depth analysis of Nietzsche's ethical thought to evaluate how his critique of Judeo-Christian morality relates to critical posthumanism's intention "not to reject humanism *tout court*—[as] there are many values and aspirations to be admired in humanism—but rather to show how those humanist aspirations are undercut by the philosophical and ethical frameworks used to conceptualize them."[14]

Nietzsche's sometimes troubling rhetoric and the uses and abuses of his philosophy by the Nazis, and today again by far-right and alt-right movements,[15] only complicate further any assessment of the ethical and political relevance of Nietzsche for contemporary posthumanism, and this explains at least in part why a number of major thinkers of posthumanism such as N. Katherine Hayles, Donna Haraway, or Wolfe either pay little attention to Nietzsche or explicitly distance themselves from his philosophy on ethical grounds. Wolfe's case is perhaps most telling in this regard. Considering how much his critical posthumanism extends poststructuralist modes of thinking that are deeply indebted to Nietzsche, the almost complete absence of Nietzsche from Wolfe's *What Is Posthumanism?* (which contains only two brief references to the German philosopher) must be surprising. Even in his essay "Human, All Too Human," with its very title borrowed from Nietzsche, Wolfe does not engage Nietzsche directly. Wolfe reveals in *Before the Law* the ethical concerns that stand behind the absence of Nietzsche from his writings. Drawing on Peter Sloterdijk's reading of Nietzsche's animal philosophy, Wolfe acknowledges that Nietzsche can help us recognize how the "cardinal biopolitical sin of contemporary practices such as factory farming . . . or routinized experimentation . . . is not just the pain and suffering it causes, . . . but rather the diminishing and deadening of 'animality' itself in all its vitality, creativity, and multiplicity, which would in turn forestall our own ability to discover the multiplicity in ourselves via animality as a creative force for our own evolution" (41). Subsequently, Wolfe nevertheless turns away from Nietzsche, citing Esposito's observation linking Nietzsche to a history of horrors that "stretches from the eugenics and selective breeding of Nietzsche's own century to the Nazi death camps" (41). In his *Bios,* however, Esposito offers a highly intricate rescue of Nietzsche from this history of horrors. He reads Nietzsche's ethopolitical writings as astute analyses of what he identifies as the immunitary paradigm central to the Western tradition of political thought. While, "with Nietzsche, the category of immunization has already been completely elaborated," (47) Esposito also suggests that, "the more Nietzsche is determined to fight the immunitary syndrome, the more he falls into the semantics of infection and contamination" (96).

This book will return to the ethopolitical dimension of Nietzsche's posthumanism throughout and specifically in chapter 7. The incompatibility of some of posthumanism's Nietzsches does not necessarily represent fissures in Nietzsche's philosophy: the "prophetic vitalist," for example, is a response to and possible remedy for the precarious insights of critical Nietzsche. Rather, the existence of so many posthumanist Nietzsches merely exposes the wide variety of interests and approaches that make up the schools of thoughts associated with posthumanism. Different strands of posthumanism read Nietzsche differently, relate to different aspects of Nietzsche's thought, and/or distance themselves from Nietzsche for different reasons. Disagreement about Nietzsche's relevance for posthumanism is further aided by the fact that posthumanists often seem preoccupied with the present and with presently imaginable futures. As a consequence, posthumanism's historical interests have been rather limited. With a few exceptions, posthumanism tends to focus on the more recent past, addressing technological and media changes of the late twentieth and early twenty-first centuries and perhaps Martin Heidegger's later work. Where it ventures further into the past, it usually does so in negative terms and with a bird's eye view, as in its often wholesale rejection of Kant and the Enlightenment.[16] And while Nietzsche is mentioned most often as an important antecedent, posthumanists do not engage Nietzsche's philosophy in a comprehensive, in-depth manner that would consider the embeddedness of his thought in the philosophical and scientific discourses of the late nineteenth century. Instead, many posthumanists make do with highly selective readings and common assumptions about Nietzsche, or rely in their assessments on Nietzsche's intellectual, cultural or (unfortunate) political legacy. This is not to downplay the significance of the historical legacy of Nietzsche for posthumanism. The period between Nietzsche's death in 1900 and the publication of N. Katherine Hayles's seminal work *How We Became Posthuman* in 1999 (which put posthumanism on the proverbial map) is full of tensions between humanist and antihumanist modes of thinking that were shaped in no small part by Nietzsche's thought. For the most part, however, it is this legacy, more so than Nietzsche's writings, that informs references to Nietzsche in the posthumanist literature. The lack of attention to the particulars of Nietzsche's arguments and their context applies as much to the problematic association of Nietzsche's overhuman with transhumanist fantasies of a genetically or technologically "enhanced" posthuman or transhuman as it pertains to critical posthumanism's indebtedness to Nietzsche via his poststructuralist reception. And it also applies to those posthumanists who distance themselves from Nietzsche's ethics based on a sample of controversial statements of his. The problem with this approach is that it lends itself, simply put, to a lot of cherry-picking and sour grapes, and thus helps expand our understanding neither of Nietzsche's philosophy nor of his continued relevance for contemporary posthumanism.

In response to the motley receptions of Nietzsche in the posthumanist literature, this book will offer an in-depth reading of his posthumanism that pays attention to the details of Nietzsche's arguments and the historical, philosophical, and scientific contexts that inform them. As indicated above, a more comprehensive assessment of the relevance of Nietzsche's philosophy for posthumanism is valuable on two accounts: First, it promises to add an important chapter to the genealogy of posthumanism while also shedding light on some of posthumanism's own internal tensions. Second, putting Nietzsche in dialogue with posthumanist theories opens up new vistas onto Nietzsche's thought. Through this dialogue, new and lesser-known aspects of his writings will receive attention. These include the neurophysiological basis of Nietzsche's epistemology, his concerns with insects and the emergent social properties they exhibit, and his reflections on the hominization effects of technology. The ethos of critical posthumanism also offers a new perspective on some of the key ethical and political contentions of Nietzsche's writings.

What (Is) Posthumanism?

Posthumanism is neither a methodology nor a philosophical school of thought, but revolves around a common theme, the redrawing and decentering of "the human." It focuses in various ways on the embeddedness of humans in material, physiological, social, and technological contexts and shares the desire to extend certain humanist values beyond the human species. Furthermore, posthumanism recognizes that, because of its anthropocentrism, the humanist tradition has not only failed to live up to many of its own lofty goals but is also unable to respond effectively to many of the challenges society faces today. As a quick look at the 2018 *Posthuman Glossary* edited by Rosi Braidotti and Maria Hlavajova reveals, these general parameters have led posthumanism to pursue a hodgepodge of often incompatible theories, concerns, topics, and ideas and equally divergent social, cultural, scientific, ethical, political, and artistic interests. A main reason for this diversity is that, unlike the theory-oriented (or theory-averse) schools of thought that dominated academia over the last fifty years, posthumanism has neither developed nor agreed to adopt a particular theoretical framework for itself. The lack of theoretical commitment is perhaps most apparent if we consider how different proponents of posthumanism position themselves vis-à-vis the main theoretical paradigms of the last decades. While some wholeheartedly endorse the heritage of poststructuralism, media theory, or cybernetics (e.g., Elaine L. Graham, Haraway, Herbrechter, Andrew Pickering, Wolfe), others seem to favor a more eclectic approach to poststructuralism (Neil Badmington, Hayles, Pramod Nayar, Tamar Sharon), while a third group—including many of the new materialists—take a directly adversarial stance toward this tradition (Braidotti, Graham Harman, Quentin Meillassoux, Timothy Morton). The lack of methodological consistency might well be viewed as strength. It increases the scope of inquiry and encourages interdisciplinarity. It also constitutes evidence of how posthumanism signals a paradigm shift, or at least creates new vantage points from where to reassess the significance of the theoretical debates of the last decades.

Drawing on Nietzsche, we can bring historical context and theoretical reflection to the many variants of posthumanism: assemblage posthumanism, critical posthumanism, dystopic posthumanism, mediated posthumanism, methodological posthumanism, panhumanism, philosophical posthumanism, radical posthumanism, and antihumanism and transhumanism or humanity+, not to mention antecedents in poststructuralist theory, early postmodernism, cybernetics, systems theory, cyborg studies, or some of the new materialisms such as Object Oriented Ontology or Speculative Realism. This book will not attempt to offer another overview of the highly diverse and fast-moving fields we associate today with posthumanism. It will also not commit to a particular strand of posthumanism, but instead draw on Nietzsche's thought as a lens to focus on the genealogy of selected posthumanist concerns, specifically on its epistemological, social, technological, and ethical contentions. Shedding light on the origins of posthumanism in nineteenth-century philosophy and the science that informed its most famous representative will allow this study both to stay attuned to conceptual challenges that run across and underneath internal divisions within the posthumanist literature and to connect Nietzsche to current debates without restricting such connections to a particular school of thought (such as the limited focus of the debates surrounding the relevance of Nietzsche's overhuman for transhumanism). Nevertheless, a few preliminary remarks about posthumanism are in order to map the terrain, narrow the field, and indicate the directions in which this book's reading of Nietzsche's posthumanism will proceed.

Phenomenology of the Posthuman

Posthumanism and "the posthuman" exist on different yet connected phenomenological planes. If we take Nayar's definition as a general guide and define posthumanism broadly as the "radical decentering of the traditional sovereign, coherent and autonomous human,"[17] then posthumanism manifests itself in three different areas: as the product of complex social, scientific, and technological developments; as an object of cultural production; and as a subject of academic and semiacademic study. As a social phenomenon, scientific and technological advances have made the "decentering" of humans increasingly palpable, and its recognition perhaps more urgent. As society is progressively digitized and computer technologies such as artificial intelligence, virtual reality, and augmented reality replace or extend what hitherto were thought to be exclusively human capabilities, awareness of the far-reaching effects of technology on what it means to be human broadens. But, as chapters 5 and 6 will argue in more detail, these are certainly not the only or first technologies that would mold what it means to be human. Communication media appear to play a crucial role here. The expanded uses of digital and social-media technologies in recent years affect how humans interact with each other and how they interface with a continually available and continually changing knowledgebase. Scientific and technological advances, whether they instill hope or fear, have also increased our awareness of environmental and other endemic and global challenges (economic, medical, political, and so on) that seem to be beyond the scale of human control. Likewise, manipulations of the body, whether through body arts, medical technologies, or chip implants, are changing the relationship between physiology, psychology, and technology. Research into animal behavior has altered and continues to alter views on the connections between humans and animals, including on shared social (or swarm) behavior. The humanities have also helped foster a new sense of ethical responsibilities, such as regarding the ills of industrialized animal farming or the bonds that can form between humans and animals—what happens "when species meet," to quote the title of Haraway's pathbreaking study from 2008.

Clearly, many of the social phenomena we associate with the advent of posthumanism are too new to be of direct concern for an inquiry into Nietzsche's posthumanism. And yet, there are relevant parallels. Nietzsche wrote in a time of immense scientific, technological, economic and political change. Like much of posthumanism, Nietzsche paid close attention to the natural sciences, in particular the life sciences, and to the precursors of today's cognitive sciences, empirical psychology and neurophysiology. His social theory, too, was informed by media technological considerations, such as the effects of the mass media of his time (the printing press). Indeed, much of Nietzsche's critique of humanist values is framed as a critique of modernity, which he sees as a period that must yet come to grips with the many changes it experiences in almost all areas of life (from the role religion plays in society to the ill effects of industrialization and the printing press, and so on). In this regard, some of the hopes and many of the fears expressed by posthumanists and transhumanists existed already in Nietzsche's time.

Pointing toward such historical continuities is not to ignore, but to argue for the necessity to examine more closely how recent advances in the sciences, how new technologies, and how the disappearance of print as the dominant communication media contribute to social changes that create the sense that the era of humanism is coming to an end.[18] Vis-à-vis these changes, Nietzsche's work encourages us to take a wider historical perspective. After all, the humanism Nietzsche targets dates back to Ancient Greece, and main tenets of its critique certainly developed before the late twentieth and early twenty-first centuries. Adopting a wider historical perspective should also help us not to get distracted too quickly by the technological glitz of today, and instead inquire how much Nietzsche and his time reacted to comprehensive social changes in which contemporary society continues to partake. If humanism is loosely identify with the age of book printing, as Herbrechter and others have suggested, Nietzsche's reflection on the ill effects of newspapers creating a new barbarism (e.g., "even great wars and revolutions are able to influence [us] for hardly more than a moment," as the "war is not even over before it is transformed into a hundred thousand printed pages and set before the tired palates of the history-hungry as the latest delicacy" [HL, 83]) might well be viewed as a recognition of a postbook culture that foreshadows some of the social effects of contemporary electronic and digital media. Without wanting to share Nietzsche's conservatism in this regard, his recognition of the effects of media both on the "message" and on social and cultural bonds suggests an important vantage point from which Nietzsche's thought signals the beginning of the end of the humanist tradition based on changes in the dominant communication media technologies.

Posthumanism manifests itself not only as a social phenomenon. The social, scientific, and technological developments associated with posthumanism are also the subjects of a broad array of cultural productions. Representations of posthumans, and to a lesser extent of posthumanist values and modes of thinking, permeate our cultural imaginary, be it in literature, film, TV, the internet, the news, or video games, or as manga, virtual worlds, or other flights of today's (often digitized) imagination. Contemporary mass media are fascinated by representations and discussions of the posthuman, for both their entertainment and their news value. Popular science stories about posthuman contentions seem particularly apt to blur the line between science and entertainment, lending the media that portray them an aura of scientificality. The ubiquitous presence of cultural and media representations of the posthuman (what Herbrechter, with reference to Slavoj Žižek's "third culture," sees as "the product of 'cognitivist popularizers' of 'hard' sciences"[19]) and "the exuberant production of ideas around the posthuman"[20] do more than simply reflect the current state of affairs; they help shape and alter perceptions of the posthuman, and by extension, of what is human.

Thirdly, then, posthumanism appears in today's society as an "-ism," manifesting itself extensively in descriptions and debates about posthumanism and its social and cultural manifestations, in discussions about the aesthetic, political, philosophical, and ethical challenges with which posthumanism confronts the so-called humanities, and in analyses of the science behind these manifestations. To no small degree, the academic literature on posthumanism has begun proliferating recursively in recent years. It is caught in a process of continued self-reflection that takes the form of overviews, introductions, handbooks, and "cartographies"[21] of posthumanism, in addition to more pointed debates about how posthumanism defines posthumanism. The stage of consolidation and self-reflection is evident in the publication of anthologies such as the 2015 *Palgrave Handbook of Posthumanism in Film and Television*, edited by Michael Hauskeller, Thomas Philbeck, and Curtis Carbonell; the 2016 *Cambridge Companion to Literature and the Posthuman*, edited by Bruce Clarke and Manuela Rossini; Braidotti and Maria Hlavajova's 2018 *Posthuman Glossary*; the 2020 *Bloomsbury Handbook of Posthumanism*, edited by Mads Rosendahl Thomsen and Jacob Wamberg, or the 2022 *Palgrave Handbook of Critical Posthumanism*, edited by Herbrechter and company. Such volumes bridge the gap between the second and third planes on which posthumanism flourishes, both as a cultural phenomenon and as an object of extensive academic and media consideration. Noting the recursiveness of posthumanist reflections on posthumanism is not meant as a point of critique. Any discipline establishes, maintains, and reproduces itself by writing about its own writings. Such self-reflections promote the formation of a discipline, as well as its evolution, by continuously negotiating and redrawing its borders.

Writings about posthumanism play a crucial role in expanding the cultural and social field of posthumanism as they venture out to discover new areas and ages of posthuman manifestations, creating an overlap between map (writings on posthumanism) and territory (the social, scientific, and cultural phenomena identified as relevant for posthumanism). This undermines any strict separation between posthumanism as, on the one hand, social condition and, on the other, something that exists discursively, a "new conceptualization of the human."[22] Instead, we need to recognize that the literature on posthumanism is always already part of the object it analyzes. Accordingly, a book on Nietzsche's posthumanism will and should also alter the understanding of posthumanism: it should help determine what phenomena are associated with posthumanism and how it is perceive and valued. Accepting the underlying circularity—that the observation affects what is being observed—has another advantage. Despite Claire Colebrook's proclamation in 2014 of the *Death of the Post-Human*, we don't have to be afraid of a future scholar, following Bruno

Latour's lead, writing a book entitled *Why We Have Never Been Posthuman*. As long as "posthumanism" exists as a concept and an evolving discursive field, posthumanism "exists," at the very least as an umbrella term that allows us to relate to each other a set of phenomena, observations, particular ways of thinking, and ethical choices.

For this book on Nietzsche's posthumanism, the most relevant aspect of the academic pursuit of posthumanism, however, is not the analysis of social phenomena or cultural representations of the posthuman. Rather, this study will follow Wolfe's understanding of posthumanism as a movement that does not announce the end of humanism *per se*, but introduces new modes of thinking that no longer take for granted as self-evident, natural, or given many of humanism's core presuppositions. For Wolfe, "the nature of thought itself must change if it is to be posthumanist."[23] Wolfe's point is ultimately political. He argues that it is not enough to decenter the human in relation to its biological, technological, or ecological position, but that a critical posthumanist approach needs to ensure that it does not "reproduce the very kind of normative subjectivity—a specific concept of the human—that grounds discrimination against nonhuman animals and the disabled in the first place."[24] This book's inquiry into Nietzsche's posthumanism follows Wolfe inasmuch as its guiding question indeed concerns how Nietzsche's writings outline new modes of thinking that challenge core propositions of the humanist tradition. Most relevant in this regard are Nietzsche's critiques of assertions of human autonomy (agency, subjectivity, rationality), of human exceptionality (within the animal kingdom or nature or as masters of technology), and of human universality (the universal applicability of certain values, aims, and goals). To understand Nietzsche's critique of these humanist conceits, this study will revisit: his critical epistemology, including his philosophy of language, by which he questions the superiority of the intellect; his philosophical anthropology, which undermines the hierarchy between humans and animals, but also between individualistic behavior and that of groups; and his recognition of the constitutive role technology plays in defining human thought and behavior. The concluding chapter will then reflect on how these topics and Nietzsche's approach to them open up new vistas on Nietzsche's and posthumanism's ethical and political contentions.

What (Is) Humanism?

A project that promises to examine Nietzsche's posthumanism cannot read the prefix "post" as marking a strict temporal border, as a simple "no more." As has been noted with regard to other "posts," the prefix equally indicates both discontinuities and continuities. Any "post" is a "post" only in relation to that from which it departs, and hence is still beheld by its other. Along these lines, Herbrechter argues that the "rhetorical essence" of the prefix "post" is that "it ambiguates. It plays with supersedence, crisis, deconstruction, regression and progression at once. Its main virtue, if one chooses to take it seriously, is to defamiliarize, detach and surprise."[25]

Acknowledging the ambiguity, and hence also continuity, of posthumanism's relationship to the humanist tradition necessitates a reflection on what is meant by "humanism." What kind of humanism and humanist modes of thinking does posthumanism (in our case, Nietzsche's posthumanism) hope to supersede? This is not a simple question. "Humanism" is a broad and quite ambivalent term, particularly when we consider its rather vague use as a general foil in most of the posthumanist literature. Furthermore, there is also the danger of conflating humanism with the Enlightenment, as the latter is often viewed as the pinnacle of the former. When Nayar defines posthumanism as the "radical decentering of the traditional sovereign, coherent and autonomous human," he clearly targets what are central tenets of the Enlightenment: its hopes to promote human autonomy, for example as an escape from what Kant defines as self-imposed immaturity (*Unmündigkeit*); its demand for political as well as moral sovereignty; and the idea that both can be accomplished through educational procedures that help us fashion a coherent and responsibly acting self.

Based on the perception that the Enlightenment extends central humanist conceits, the Enlightenment is also often used to delineate different strands of posthumanist thinking, revealing deep fault lines within this vast and unwieldy field of inquiry. Critical posthumanism in particular uses the Enlightenment as a measuring stick to differentiate itself from strands of posthumanism that it criticizes for extending rather than refuting major tenets of Enlightenment thought. In *How We Became*

Posthuman, Hayles already accused Hans Moravec of "not abandoning the autonomous liberal subject but expanding its prerogatives into the realm of the posthuman."[26] Wolfe singled out Bostrom's transhumanism/ humanity+ as a mode of thinking that "derives from the Enlightenment belief in human perfectibility."[27] Natasha Vita-More's "Transhumanist Manifesto" acknowledges as much: "Transhumanism offers a new philosophical approach to the human condition while simultaneously expanding upon antecedents—the Renaissance, Enlightenment, Modernism, and Postmodernism."[28] Popular futurists like Ray Kurzweil or Francis Fukuyama are equally unabashed about their allegiance to Enlightenment rationality and values and its anthropocentrism. Herbrechter, whose *critical* posthumanism itself extends a central Enlightenment conceit, finds the Enlightenment ethos continued in the antihumanism of Nietzsche. The "last and most radical inheritor of Enlightenment thought," Herbrechter argues, still pursues a hermeneutics of suspicion "aimed at a renewal and a liberation of 'man' from 'his self-imposed Immaturity.'"[29]

Considering posthumanism's as well as Nietzsche's complicated relationship to the Enlightenment (is Nietzsche the Enlightenment's last and most radical inheritor or is he its first posthumanist critic?), it is important that we more clearly delineate humanism from the Enlightenment. Michel Foucault had warned already against blanket identifications of humanism with Enlightenment thought. In his essay "What is Enlightenment?," Foucault argues that "just as we must free ourselves from the intellectual blackmail of being for or against the Enlightenment we must escape from the historical and moral confusionism that mixes the theme of humanism with the question of the Enlightenment" (45). Foucault understands the Enlightenment along the lines set out by Kant, as a continued, open-ended project of critical (self-)examination, arguing that what connects us "with the Enlightenment is not faithfulness to doctrinal elements, but rather the permanent reactivation of an attitude— that is, of a philosophical ethos that could be described as a permanent critique of our historical era" (42). In this respect, the Enlightenment is in principle opposed to the preservation of values and norms inherent to the wide variety of past and present humanisms. Foucault indeed offers a catalogue of humanisms that the Enlightenment, as a critical discursive practice, would have to oppose: Christianity and other religions, divergent nineteenth-century strands with a "suspicious humanism hostile and critical toward science and another that placed its hope in that same science" (44), and the humanist contentions of Marxism, which Foucault juxtaposes to those of National Socialism, before pointing out that even some Stalinists "said they were humanists" (44). The list emphasizes the normative tendencies of humanism and shows how various notions of "the human" have been employed in support of vastly differing ideologies where humanism serves "to color and to justify the conceptions of man [borrowed from religion, science, and politics] to which it is after all obliged to take recourse" (44).

For Foucault, the critical impetus of the Enlightenment serves as an important antidote against the normative force of the humanist traditions. The tendency to employ humanism for biopolitical purposes "can be opposed by the principle of a critique and a permanent creation of ourselves in our autonomy: that is a principle that is at the heart of the historical consciousness that the Enlightenment has of itself" (44). Let's note, then, that, as a method of permanent critique, the Enlightenment is *not* opposed to but is a crucial component of a *critical* posthumanism deserving of this name. Indeed, when Braidotti concludes the introduction to her 2013 book *The Posthuman* with the promise that the "posthuman condition urges us to think critically and creatively about who and what we are actually in the process of becoming,"[30] she places the process of becoming posthuman squarely within the historical consciousness of the Enlightenment as described by Foucault. And it is in the same spirit that Luhmann, whom Wolfe considers along with Jacques Derrida to be an "exemplary posthumanist theorist,"[31] published many of his most important essays in a six-volume series of books he entitled *Sociological Enlightenment*.

With regard to Nietzsche's posthumanism, the task of delineating what kind of humanism and humanist modes of thinking he targets is further complicated, but perhaps also narrowed, if Enlightenment thought is not subsumed under the umbrella of humanism. Nietzsche's appraisal of the history of Western metaphysics includes a critique of Enlightenment assumptions about the centrality of reason and the human intellect and its belief in progress, but historically he traces these aspects of humanist thinking back much further, to Socrates, Euripides, and Plato, and to the neo-Platonism of Judeo-Christian religious traditions. According to Nietzsche, it is in his time, in the second half of the nineteenth century, and because of the pressure of continued critical and self-critical modes of thinking, including in the natural sciences, that the humanist paradigm of thinking was coming to a head.

To profile the humanism that is the target of Nietzsche's critique and that is distinct from the critical consciousness of the Enlightenment, I want to turn to Heidegger's *Letter on Humanism,* a text that offers an expansive definition of humanism that aligns historically with Nietzsche's critique of the Western tradition, while also explicitly defending a posthumanist ethos. Originally written in 1946 to Jean Beaufret and revised for publication in 1947, in the immediate aftermath of earth's greatest human-made humanitarian catastrophe, Heidegger's letter addresses the question of how humanist modes of thinking are linked to the inhumaneness that history continues to witness and that had escalated in the twentieth century. In *Platons Lehre von der Wahrheit* (Plato's Doctrine of Truth), which he published alongside his *Letter on Humanism,* Heidegger argues that the "beginning of metaphysics in Plato's thinking is at the same time the beginning of 'humanism.'"[32] He expands on this point in the *Letter on Humanism.* While historically humanism is first explicitly considered and sought as an ideal in ancient Rome, the Romans incorporate and extend the self-understanding of the ancient Greeks when they define themselves as *homo humanus* in contradistinction to *homo barbarus.*[33] The Italian Renaissance of the fourteenth and fifteenth centuries, the age of humanism in the narrower sense, is a *renascentia romanitatis* in which the "in-humane" against which humanism defines itself has been replaced by the "barbarism of the gothic scholasticism of the Middle Ages" (225). The ancient roots of humanism are evident still in the eighteenth century, where the humanism associated with Weimar Classicism—Heidegger mentions Johann Joachim Winckelmann, Johann Wolfgang von Goethe, and Friedrich Schiller, but exempts Friedrich Hölderlin—goes hand in hand with a revival of ancient Greek culture. When Heidegger notes toward the end of *Plato's Doctrine of Truth* that, in Nietzsche's thinking, the history of metaphysics as it originated in Plato's thought "entered upon its unconditioned fulfillment,"[34] he suggests that Nietzsche's writings are still part of, but also mark the end of humanism.

Heidegger's critique of humanism is not based on rejecting its anthropocentrism altogether, or its hyperbolic notions of autonomy, sovereignty, and coherence. Rather, the problem lies with humanism's failure to hold humans in high enough esteem. In the *Letter on Humanism*, he argues that the highest determinations of the essence of humans in humanism, "the humanistic interpretations of man as *animal rationale*, as 'person,' as spiritual-ensouled-bodily being," do not realize the "proper dignity of man" (233). To grasp this failure, it is necessary to understand how Heidegger construes the essence of humanism. Heidegger defines humanism as the "meditating and caring, that man be human and not in-humane, 'inhuman,' that is, outside his essence" (224). This seemingly tautological definition leaves open, and hence casts as malleable, what defines the humaneness and humanness of human beings. Heidegger references Karl Marx, to whom humans are most naturally human in their social existence, as society secures equally the nature and natural needs of human beings. He juxtaposes Marx's socially defined essence of humanity to the Christian view. Christianity delineates "the humanity of man, the *humanitas* of *homo*, in contradistinction to *Deitas*" (224). The Christian perspective reveals the Platonic origins of humanism, echoing a central tenet of Nietzsche's critique of Christianity, as it insinuates that "man is not of this world, since the 'world,' thought in terms of Platonic theory, is only a temporary passage to the beyond" (224).

It should not be ignored that Heidegger's philosophy is tainted by his initial support for the Nazis and, as was discovered more recently, that he had privately shared some of their anti-Semitism. The problem, however, lies not with Heidegger's analysis of humanism, but with the fact that important aspects of his thinking do not escape the metaphysical presuppositions that he locates at the core of humanism and blames for its failure to think the dignity of humanity high enough. This is a point Theodor Adorno makes with regard to the elitism he detects in Heidegger's "jargon of authenticity."[35] It is flushed out more fully recently by Jean-Luc Nancy, who is able to link Heidegger's troubling political stance to an underlying contradiction of his ontology which "left a place—and not the least important—for a decisive element of the metaphysics of being: the presupposition of the initial, of the foundation and the origin, of the authentic and the proper."[36] It is because of his own metaphysics of being that Heidegger fails, we might add, to think the dignity of humanity high enough. In this regard, his *Letter on Humanism* raises a question that is also central to Nietzsche's political legacy: whether the inhumaneness we continue to witness and that escalated in the twentieth century is at least in part generated by humanist modes of thinking and valuing, or whether the forces opposed to this tradition are responsible for the continued inability of society to overcome its violent tendencies. In other words, are Nietzsche and Heidegger and their critiques of humanism part of the problem, or are they analyzing problems at the heart of the humanist tradition in search for posthumanist alternatives (in values and thinking) that would offer the outlook of a more humane society?

This book supports the latter position. It will address some of the ethical questions raised by Nietzsche and posthumanism at various points and focus more extensively on the ethical and political implications of his posthumanism in chapter 7. For the discussion at hand, Heidegger's association of humanism with Platonic metaphysics is relevant because it makes it possible to identify the problematic core shared by diverse humanist traditions without having to reject blankly all their values and anthropocentric conceits. This is particularly true for the Enlightenment, where it would in fact be a sign of immaturity if, with the bathwater, posthumanists also wanted to throw out such precious values as the ideals of humaneness, justice, equality, civility, and peace propagated by its leading figures. As Wolfe argued, critical posthumanism's intention is "not to reject humanism *tout court*—indeed there are many values and aspirations to be admired in humanism—but rather to show how those humanist aspirations are undercut by the philosophical and ethical frameworks used to conceptualize them."[37] Locating the core of humanism in its adoption of Platonic metaphysics makes it possible to be more strategic about the values and norms set forth by the Enlightenment. We can thus avoid a stance that merely cherry picks, as Colebrook laments, certain enlightenment values over others ("Yes, we want the rights and freedoms of the enlightenment but should be wary of universalizing specifically modern Western values"[38]), but continues to affirm a "thermodynamic" rationality, and with it a "cosmopolitanism" that fails to change the presumptuousness of the Enlightenment. Viewed through the lens of Heidegger's critique of humanism, the light is put on onto-theological binaries that are also the target of Nietzsche's critique. Adopting this lens, it is possible to critique humanist modes of thinking while continuing to support values and ideals that "at times," as Jane Bennett observes, "worked to prevent or ameliorate human suffering and to promote human happiness or well-being."[39]

The brief genealogy of humanism Heidegger developed in his *Letter on Humanism* submits that the essence of humanism does not lie in any particular qualities that are attributed to human nature or human (moral) behavior—those are subject to historical circumstance and change—but hinges on how humanism delineates itself against a barbaric Other, as well as on humanism's adoption of a Platonic mode of thinking that privileges a realm of the ideal over a worldliness that is viewed as shadowy, transient, deficient, and so on. From its origins in ancient Greece's understanding of its civilization in counterdistinction to the noncivilized, barbarian world through the adoption and adaption of this distinction by the Romans, in the Renaissance, and into Weimar Classicism, humanism has retained its coherence and prerogatives by defining the essence of the human and what it considers humane in counterdistinction to an Other. Its self-determination qua delineation from a barbaric Other is a structural property of humanism. It persists throughout the ages Heidegger references and carries within it the seed of its failure, its inability to live up to its lofty goals. It is also present in the way humanism adopts Plato's idealism. In particular, in its Christian variant, humanism's discriminatory rationale becomes part of its basic anthropology. It delineates the essence of the human(e) from its material, physiological, and "animal" existence, devaluing the latter as transitory, nonessential, contingent, and inherently sinful and guilty. In the Judeo-Christian tradition (which for Nietzsche as well as for Heidegger extends into Enlightenment thought), the barbaric Other is what is perceived as the animal part of the *animal rationale*. It is, in other words, Platonic and neo-Platonic idealism that underwrite the chastising of humanity's physiological and social existence, of drives, desires, wants, and ultimately of "life" as Nietzsche understands it.

The problem Heidegger identifies in this mode of humanist thinking, why it must fail "to realize the proper dignity of man," then, is twofold. Through the process of delineation, the Other continues to inform the definition of that which is cast as human(e). This logic of inclusion qua exclusion extends specifically to the religious separation of the human(e) humans from the animal human. In the *Letter on Humanism*, Heidegger points out that, by continuing to delineate humankind as living being different from plants, animals, and God,

> we abandon man to the essential realm of *animalitas* even if we do not equate him with beasts but attribute a specific difference to him. In principle we are still thinking of *homo animalis*—even when *anima* [soul] is posited as *animus sive mens* [spirit or mind], and this in turn is later posited as subject, person, or spirit [*Geist*]. Such positing is the manner of metaphysics. But then the essence of man is too little heeded and not thought in its origin, the essential provenance that is always the essential future for historical mankind. Metaphysics thinks of man on the basis of *animalitas* and does not think in the direction of his *humanitas*. (227)

Combined, these modes of thinking undercut "the concern that man become free for his humanity and find his worth in it" (225).[40]

Heidegger's *Letter on Humanism* offers a background against which we can assess Nietzsche's posthumanism, not only in terms of the historical narrative Heidegger lays out or the problematic (bivalent) logic he detects at the center of the humanist enterprise, but also in terms of the ethical conclusions he draws, with the hope that it will become "somewhat clearer . . . that opposition to 'humanism' in no way implies a defense of the inhuman but rather opens other vistas" (227). The possibility of such new vistas becomes more apparent if we clearly differentiate the Enlightenment as a critical project from the normative force of a broadly defined humanism grounded in neo-Platonic metaphysics. By doing so, a critical posthumanism need not reject ideas of progress, though progress should not be hypostasized as an independent force, but measured against clearly defined parameters. Nor does a critical posthumanism have to jettison the idea that its ethos could and should aim toward a betterment of the human species, a betterment that, put bluntly, is not contradicted, but aided by its concern for animal rights and the environment. Critical posthumanism instead must target what is also at the center of Nietzsche's critique of Western thought: the Platonic and neo-Platonic idealism that places a human essence above and beyond everything else, as an autonomous agent that is thought to exert cognitive and moral authority over its Others.

Posthumanism's Great Divide

Heidegger's understanding of humanism also makes clear the philosophical depth associated with any posthumanism that hopes to challenge humanist modes of thinking, including humanism's grounding in a certain (Platonic) epistemology. It is not coincidental that Heidegger published the *Letter on Humanism* alongside his essay on *Plato's Doctrine of Truth*, a text that, based on Nietzsche's deconstruction of a metaphysical conception of truth, credits Nietzsche's work as marking the beginning of the end of the metaphysical tradition of thought Heidegger sees as the foundation of humanism. In this regard, any inquiry into Nietzsche's posthumanism will have to revisit his critical epistemology—which is the task of the next chapter. Before turning to Nietzsche's epistemology, however, I want to conclude this chapter reflecting on the great divide that separates and makes it possible to group different schools of posthumanist thought. This divide is most apparent in the above-mentioned rejection of transhumanism by critical posthumanism. While critical posthumanism seeks to break with humanism's long-standing metaphysical contentions, transhumanism, as Elaine Graham puts it, continues to reflect "a deep-rooted philosophical tradition in Western thought, a Platonic worldview in which the physical sensory world is but a reflection of a purer, ideal realm of perfect form."[41] As indicated above, this divide has led to the adoption of quite different Nietzsches by posthumanists of different stripes.

One way of looking at the divide between transhumanism and critical posthumanism, and now the new materialisms, is their different stances toward poststructuralism, which range from ignoring, to endorsing, to outright rejecting its basic lessons. Sharon recently offered a more refined cartography of posthumanism. In her 2014 *Human Nature in an Age of Biotechnology,* she makes posthumanism's stance toward conceptions of human nature the centerpiece of her "cartography of the posthuman" (the title of her chapter 2). Sharon distinguishes between, on the one side, a dystopic posthumanism and a liberal posthumanism and, on the other, a radical and a methodological posthumanism (and adds as a fifth option, her own "mediated posthumanism").[42] She profiles the rifts in terms of their different attitudes toward the notion of human nature. Dystopic and liberal posthumanism defend notions of human nature, albeit different ones: the former sees technological and biotechnological advances as a threat, while the latter posits that the same advances have the potential to heal, complete, and expand human capabilities. According to this taxonomy, transhumanism is part of "liberal posthumanism," which "is characterized by an endorsement of emerging biotechnologies for their perceived ability to allow humans to transcend their biological limits and enhance themselves at will."[43] On the other side, radical and methodological posthumanism share their opposition to such essentializations of human nature. Radical posthumanism, which aligns by and large with Herbrechter's critical posthumanism,[44] neither defends nor wants to extend human nature, but targets instead the constraints and hierarchies that come with the former's (and humanism's) subjection to an essentialized notion of human nature. Likewise, methodological posthumanism does not rely on an essentialization of human nature, but it lacks the political impetus of radical posthumanism. Sharon includes science and technology studies (STS) and "actor-network theory" (ANT) scholars like Michel Callon and John Law, Bruno Latour, and Andrew Pickering, as well as the newer generation of philosophers of technology such as Don Ihde and Peter-Paul Verbeek, in this group. I would add neocybernetic and systems-theoretical thinkers like Gregory Bateson, Niklas Luhmann, Dirk Baecker, and Cary Wolfe to this group. They contribute to a methodological posthumanism that aims "to conceptualize analytical frameworks that can better account for the networks and zones of intersection between the human and the non-human."[45]

The theoretical landscape is of course more complex than any summary of it. There are plenty of posthumanists who straddle the line or cross back and forth between different strands of posthumanism or find themselves on both sides of the divide.[46] Yet, the discussion of this rift has remained a salient feature of writings on posthumanism. Sharon's cartography is useful for this book not just as an overview of different groupings with often starkly different outlooks within the field of posthumanism, but also because it makes it possible to situate more clearly Nietzsche's posthumanism. Sharon mentions Nietzsche several times in passing, recognizing his significance for Foucault and for Gilles Deleuze and Félix Guattari, who are "naturally indebted to Nietzsche, for whom the body is a composition of forces and should be understood in terms of quantities and qualities of forces."[47] That is, she recognizes Nietzsche as a progenitor of radical posthumanism, which "views the posthuman as providing a means of political resistance against the metanarratives of modernity and as having the potential to usher in a postmodern and post-anthropocentric era."[48] The association of Nietzsche with radical posthumanism is certainly not surprising, considering Nietzsche's importance for poststructuralist and early postmodern theory. It nevertheless bears underlining, as the most extensive discussions of Nietzsche's posthumanism have taken place on the other side of the onto-epistemological divide—that is, on the side of transhumanist appropriations of Nietzsche the "prophetic vitalist." While agreeing with Sharon's association of Nietzsche with what she calls radical posthumanism, this book's focus on Nietzsche's scientific and entomological studies, as well as his engagement with technology, aims to make the case that, in addition to his "radical" inclinations, Nietzsche must also be viewed as working toward a methodological posthumanism, toward, in Sharon's words, the conceptualization of "analytical frameworks that can better account for the networks and zones of intersection between the human and the non-human."[49] Specifically, this book contends that Nietzsche's philosophy's indebtedness to the hard sciences has him work already toward an episteme that recognizes, as Sharon notes with reference to Latour's work, how "the prevalence of human/non-human couplings and networks indicates that humans do not necessarily have a monopoly on agency, intentionality or morality, which can be extended to artifacts, as something that is 'delegated' to them, or inherently theirs."[50]

Sharon's cartography is also useful because it shows how fundamentally differing epistemologies are underlying posthumanism's great divide and its conflicting stances toward human nature. To put it bluntly, theories that essentialize human nature (or the material for that matter) subscribe to the classic idea that the intellect is capable of adequately representing the things "out there" (that are thought to exist independent of the means of their observation), while those who reject the essentialization of human nature build on the theoretical insights of poststructuralism or on constructivist epistemologies such as Latour's ANT, neocybernetics, or systems theory that maintain, simply put, the constitutive role of the means of observation in what is being observed. Revisiting Nietzsche's epistemology and its foundations in nineteenth-century biological and neurophysiological science (precursors to today's cognitive sciences) can add historical context as well as conceptual precision to posthumanism's great divide, adding perspective particularly to continued discussions of embodiment and embodied cognition and with regard to more recent attempts by Object-Oriented Ontology and other speculative realisms to circumvent the (post-Kantian) philosophical and neurophysiological insights of the nineteenth century in their quest to return to the "in-itself."

It is against the backdrop of these comprehensive posthumanist concerns that the following chapter will return to Nietzsche's early epistemology and its foundations in the physiological and neurophysiological research of his time. The late eighteenth and the nineteenth centuries are filled with discussions of mind–matter and mind–body dualisms and include plenty of reflection on phenomena that fit under the category of what Hayles, in her more recent work, calls the "cognitive nonconscious."[51] Merely think of Kantian schemata, discussions of apperception in German Idealist philosophy (especially Kant and Fichte), or of Schopenhauer's nonconscious world of the will. Just as it happens for today's posthumanists, it is with respect to the scientific literature of his day that Nietzsche's epistemology forms. And it is because of their scientific foundations, which also constitute the foundations of today's cognitive sciences, that Nietzsche's epistemological reflections remain relevant for contemporary debates on embodiment as much as they remain relevant for a critical assessment of some of posthumanism's materialist contentions.

2 Posthumanist Epistemology

> What will mankind have come to know at the end of all their knowledge?—their organs!
>
> —Friedrich Nietzsche, *Daybreak*

AT THE BEGINNING OF THE TWENTY-FIRST CENTURY, it seemed that much of the humanities had accepted the Nietzschean equation of truth with power and the recognition of the constitutive role of language in creating sociopolitically consequential realities. Along with Nietzsche's poststructuralist reception, it was Michel Foucault's work that was most influential in promoting this Nietzschean view across disciplines in the humanities. It infused gender, LGBTQ+, and ethnic studies with powerful arguments that have led to a far-reaching redrawing of the borders of traditional conceptions of "the human." In recent years, the idea that gender, race, and other forms of self-identifications are not facts of nature but constructs that have been used and abused for political purposes and that are open to change has also entered the public perception more widely. In my own experience as a professor at a midwestern state university, teaching Judith Butler's concept of gender only a decade ago was still a challenge reserved for upper-level and graduate classes. Today, my freshmen honors students rarely blink anymore when introduced to the idea of gender as a performance and as something that, within a scene of constraint, is "capable of being constituted differently."[1] Though not without setbacks, the success of this development is also apparent in the political and legal changes that have accompanied the rapid changes in perceptions of gender and sexual orientation in the last two decades.

Despite the institutional and cultural successes of what is part of

Nietzsche's poststructuralist legacy, the underlying epistemological contentions have come under scrutiny again in recent years. The opposition against constructivist notions of truth this time around arises not so much from traditionalists who hope to hold on to representational modes of thinking and firm norms, but is driven by the troubling consequences of a "post-truth" rhetoric invading politics and by new materialists who hope to undo the lessons of Kant's epistemology (on which Nietzsche built his) in their pursuit of an object-oriented ontology or speculative realism. The hope of these strands of posthumanism is to evade Kant's "correlationism" and find access again to a nonanthropomorphic, nonrhetorical reality that is no longer merely phenomenological.

If the poststructuralist extension of Nietzsche's epistemology is politically interested, so too is the new materialisms' rejection of Kant's correlationism, albeit in a different direction. The former recognizes Nietzsche's equation of truth with power for its potential to effect "good trouble" in the social realm by challenging existing norms and their discriminatory consequences. To do so seemed especially urgent in the aftermath of an era that had just witnessed the most egregious atrocities committed in the name of supposed "truths" about race, religion, and sexual orientation. The new materialists respond to a different challenge: the material and environmental effects that humanity in the age of the Anthropocene has brought upon this earth, and by extension upon itself. Epistemologically at stake here is less how humans perceive each other, and more what seems to elude immediate perception and thus prevents necessary action or the modification of existing behavioral patterns (such as the burning of fossil fuels). Timothy Morton has argued forcefully that even the most basic empiricist propositions fail to help us address pressing environmental questions. In a nutshell, asserting that a falling tree produces a sound only in the presence of an ear does little to help us address the comprehensive, often imperceptible effects of environmental pollution and climate change that make more and more trees fall. To mention another example of the posthumanist trend away from post-Kantian and poststructuralist epistemological sensibilities, Rosi Braidotti rejects the "moral and cognitive relativism" she associates with "postmodernist deconstructions" and advocates instead for posthuman research that "is neo-foundationalist and aims at re-grounding concepts and practices of subjectivity in a world fraught with contradictory socio-economic developments and major internal fractures."[2]

This chapter will revisit Nietzsche's epistemology—which is decidedly neo-Kantian—and read his linguistically motivated equation of truth with power against the backdrop of the neurophysiological science of his time with which he was familiar. Understanding the science Nietzsche consulted makes apparent that his epistemology anticipates not just poststructuralist, but also certain cybernetic contentions about cognition and communication. The latter allows us to appreciate the connection Nietzsche draws between truth and power so central to his poststructuralist reception, while also shining a critical light on the critique of Kantian epistemology by some of the new materialists without having to jettison the idea that there are realities that exist independently of the realm of immediate sensory perception and that call for urgent political action. In fact, this chapter contends that critical and methodological strands of posthumanism are positioned better to recognize and address environmental threats than is the wholesale rejection of so-called Kantian correlationism.

The Principle of Specific Nerve Energies

As early as 1862, we find references in Nietzsche's notebooks that reflect his interest in sensory perception, physiology, and neurophysiology. By 1868, as Nietzsche was working on a never-completed dissertation with the preliminary title *"Der Begriff des Organischen seit Kant"* (The concept of the organic since Kant), he was familiar with the works and ideas of scientists such as Paul Heinrich Dietrich Holbach, Rudolf Virchow, Rudolf Hermann Lotze, Hermann von Helmholtz, Matthias Jacob Schleiden, Jakob Moleschott, and most importantly for the purposes of this book, Johannes Peter Müller.[3] Christian Emden's 2005 study *Nietzsche on Language, Consciousness, and the Body* discusses in more detail the long list of books on physiology, organic electricity, and psychophysics that the young Nietzsche read or had indirect knowledge of through lectures and secondary literature, arguing that any "serious examination of Nietzsche's understanding of language and thought must . . . focus on the ways in which his notion of metaphor is connected to the intellectual fields and epistemic transitions of his time."[4] In the early 1870s, Nietzsche also studied intensively (and later on repeatedly returned to) Friedrich Albert Lange's monumental *History of Materialism,* which contains extensive discussions of contemporary science and their philosophical significance.

From the plethora of materials, I want to focus on one specific influence on Nietzsche, one Emden pays less attention to, which is Johannes Peter Müller's so-called principle of specific nerve energies (*Gesetz der spezifischen Sinnesenergien*). Müller was a towering figure in the nineteenth-century life sciences. Among his students and coworkers he counted renowned scientists such as Helmholtz, Virchow, Emil du Bois-Reymond, Ernst Haeckel, Friedrich Gustav Jakob Henle, Theodor Schwann, and Wilhelm Wundt. His lab work and experiments, including on himself, were visionary, and his two-volume *Handbuch der Physiologie des Menschen für Vorlesungen* (1833–1840; published in translation as *Elements of Physiology*) went through multiple editions in the nineteenth century.[5]

Müller's most radical and, for our purposes, most important insight was his so-called principle of specific nerve energies, which he first articulated in his pathbreaking 1826 treatise *Zur vergleichenden Physiologie des Gesichtssinnes des Menschen und der Thiere* (On the comparative study of the physiology of vision in humans and animals). He discussed the principle further in *Über die phantastischen Gesichtserscheinungen* (On imagined visual representations), a text that was published the same year and that he calls the continuation of the *Zur vergleichenden Physiologie des Gesichtssinnes.* Explanations of the principle of specific nerve energies were also included in the widely distributed *Handbuch der Physiologie.* The principle challenges the understanding of the sensory apparatus as a passive conductor of information it receives from the outside. Müller notes that his experiments show that "all sensory nerves are receptive to electricity, for example, and yet each sense reacts differently to the same cause [*Ursache*], one nerve sees light from it, the other hears a sound, the other smells, the other tastes electricity, the other feels it as pain or pressure."[6] Likewise, the nerve cells produce always the same nerve-specific output, independent of the quality of the source of their stimulation. Müller mentions specifically the eye as an example. "It does not matter what irritates the eye, it may be pushed, pulled, pressed, or galvanized, or it might sense the irritations transmitted sympathetically from other organs; for all these different causes, ... the light nerve will sense its affection as sensation of light, and sense itself as dark when resting."[7]

The consequences of Müller's findings are far reaching. They demonstrate that the output our senses produce does not reflect anything qualitative or "essential" about the input that the nerves process, but reflect merely the specific capabilities of the human physiology. This aspect is inherent to the concept of "specific energies," which in the early nineteenth century commonly referred to structuring properties, to a form of energy that should not be confused with the physical concept of (constant) energy that informs the second law of thermodynamics as conceived by Müller's most famous student, Helmholtz.[8] The conclusion Müller draws is unequivocal: in our sensory perception of the world, we constantly sense *ourselves*, the capabilities of our nerves to react to stimuli, without these sensations reflecting the nature or qualities of the things we perceive. Only after the fact is the sensation externalized "as a consequence of the interaction between the idea [*Vorstellung*] and the nerves, not of the sense alone which isolated would only sense its affections."[9] In a strict sense, then, we cannot even trust our senses to recognize the location of the source of stimulation to distinguish clearly between inner and outer stimuli. For Müller, it is the result of habitual use that leads us (without us being aware of it) to externalize what we hear and see. We are more willing to accept that feelings such as pain or happiness represent inner states rather than outer qualities, but from a sensory perspective, seeing and hearing are the same as feeling emotions: they measure inner states; they reflect the qualities of the perceiving organ rather than represent, copy, or express qualities of an outside.

Jonathan Crary points out some of the "nihilistic" consequences of Müller's theorem: it conceives of the relationship between stimulus and sensation as fundamentally arbitrary; it drains interiority "of any meaning that it had for a classical observer," stipulating instead that "all sensory experience occurs on a single immanent plane" and that it dissolves the unity of the subject into "a composite structure on which a wide range of techniques and forces could produce or stimulate manifold experiences that are all equally 'reality.'"[10] In Crary's assessment, these consequences mark Müller's principle of specific nerve energies as one of the most controversial ways in which the subject was figured in the nineteenth century.

Müller's law did not remain unchallenged in the nineteenth century. Only a few years after its publication, Lotze, a materialist biologist, contested the "specificity" attributed to nerve cells, as his research showed how nerves can acquire different functions. Lotze also separated nerve stimuli from the "soul" more clearly than did Müller, arguing that the qualities of sensations are "a production of the soul according to its laws, and do not depend at all on the nature of the physical stimuli except that these offer signals for the soul to create ideas."[11] Furthermore, as William Woodward notes, in Lotze "energies" gave way "to a proportionality between three disparate processes: the stimulus, the nervous excitation, and the conscious sensation."[12] The critique of Müller's contemporaries and subsequent modifications to the law (including Helmholtz replacing specific sensory pathways with specific fibers)[13] do not, however, take away from Müller's contribution to an episteme that has led today's neurosciences to understand consciousness and cognition as emergent phenomena tied to the dynamics of self-referencing neural networks and reciprocally interacting systems. With his law of specific nerve energies, Müller anticipated what Evan Thompson and Francisco Varela describe as "enactive" or "radical-embodiment,"[14] a conception of cognition that recognizes how "our environment emerges through embodied interactions that create meaning and make sense of otherwise arbitrary sensory input."[15] This research tradition offers an alternative to reductionist and materialist approaches that came to dominate the life sciences already in Müller's lifetime. They also escape, as I hope to show in the following, the new materialisms' accusation of correlationism.

Nietzsche's Physiology of Cognition

There is ample evidence that Nietzsche was familiar with what was Müller's most important contribution to nineteenth-century physiology.

In the spring of 1868, while working on *The Concept of the Organic since Kant,* Nietzsche put a number of nineteenth-century physiologists, including Müller and Helmholtz, on his to-read list. Helmholtz was the most famous of the long list of prominent students taught by Müller. Differences between student and teacher aside,[16] Müller's principle of specific nerve energies forms the acknowledged foundation of Helmholtz's optics. It is not surprising, then, that Helmholtz's *On the Sensation of Tone as a Physiological Basis for the Theory of Music* from 1863—which we know Nietzsche read[17]—starts by summarizing Müller's theorem. Mirroring closely the wording from Müller's *Handbuch der Physiologie,* Helmholtz writes in the third sentence of the first paragraph of the book: "Each sensory organ communicates [*vermittelt*] specific sensations that cannot be excited by another; the eye sensations of light, the ear sensations of sound, the skin tactile sensations."[18]

The clearest evidence of Nietzsche's familiarity with the principle of specific nerve energies, however, can be found in the 1872–1873 essay fragment *On Truth and Lies in a Nonmoral Sense.* In the essay, the neurophysiological vocabulary is introduced in the third paragraph. As Nietzsche expands on the limited worth of the intellect, he first turns to the illusionary quality of cognition, likening sense perception to a tactile game that is played on the back of things:

> [Humans] are deeply immersed in illusions and in dream images; their eyes merely glide over the surface of things and see "forms." Their senses nowhere lead to truth; on the contrary, they are content to receive stimuli and, as it were, to engage in a tactile game [*tastendes Spiel*] on the backs of things. (*TL*, 80)

Illusion, it should be noted from the outset, is not thought in opposition to cognition. Rather, Nietzsche's claim is that consciousness, what Arthur Schopenhauer calls the world of representation that encompasses perception, cognition, and imagination, constitutes a sea of illusions and dream images. Simply put, Schopenhauer's distinction between the world of representation and the world of the will challenges the Enlightenment's reliance on cognition, arguing that, precisely where humans perceive, recognize, or understand, they do *not* penetrate the essence of things, but rather produce images that are necessarily different from, cover up, and make inaccessible what is stipulated to be the underlying reality, the world of the will.

Chapter 4 will return to Nietzsche's conception of the will more extensively. Important for a consideration of his epistemology is how Nietzsche retains the separation between illusionary representations, on the one hand, and withdrawn "things," on the other, but already tweaks the customary relationship between the two. Describing the relationship between the two as a playful feeling around, he invokes blindness, challenging the metaphorical reliance on vision in traditional epistemologies; he also understands sensory perception (*Empfindung*) not receptively, but as an exploratory activity, as an appetitive gathering of stimuli on the back of things.[19] Furthermore, Nietzsche conceives cognitive activity as something playful, suggesting that cognition develops and follows its own rules. The break with a representational paradigm of cognition becomes still more apparent if we interpret the "tactile game" as a reference to piano playing (Nietzsche uses the verb *tasten* which, as a noun, is also the word for piano keys in German). According to the piano metaphor, the phenomenological world, which appears and takes form as a result of playing keys, is the product of the instrument that is being played. Such a reading is in tune with neurophysiological models of cognition that understand nerves no longer as passive conductors (the model Müller rejected as outdated),[20] but find the phenomenological world to be the product of the inner dynamics of the organism.

Nietzsche subsequently includes language in his physiological reflections. After having noted the arbitrariness of words and how they constitute generalizations that make us overlook what is actual and individual, Nietzsche draws on the neurophysiological discourse to emphasize how words are irreparably severed from the outside they are supposed to describe:

> If [man] will not be satisfied with truth in the form of tautology, that is to say, if he will not be content with empty husks, then he will always exchange truths for illusions. What is a word? It is the copy in sound of a nerve stimulus. But the further inference from the nerve

stimulus to a cause outside of us is already the result of a false and unjustifiable application of the principle of sufficient reason. (TL, 81)[21]

While Schopenhauer was aware of Müller's work and also refuted a correspondence between object and subject,[22] Nietzsche recognizes the still more radical consequence that follows from Müller's principle. If a nerve cell transmits no information about the source of its stimulation (it can be artificially stimulated or "fire" on its own), physiology does not provide us with a firm ground to distinguish between inner and outer, and therefore denies us the space to introduce a causal relationship between them.[23] By claiming a misapplication of the principle of sufficient cause, Nietzsche ultimately targets the role and reliability Schopenhauer attributes to the intellect in its construction of the world of representation. Nietzsche thus moves beyond Schopenhauer on two accounts. He questions Schopenhauer's reliance on the primacy of causality (a skepticism Nietzsche retains throughout his work), and he dispenses of a notion of the intellect as a central, unifying agent where the nerve stimuli would be synthesized. Nietzsche subsequently insists on separate observational planes, perhaps one of the most far-reaching consequences he draws from Müller's principle.

Nietzsche begins to dissemble the unity of the subject Schopenhauer sees guaranteed by reason when he hones in on the relationship between nerve stimuli and language in the immediately following paragraph, where he notes that the separation between mental image and language is the same as that between mental image and nerve stimulus:

> The "thing in itself" (which is precisely what the pure truth, apart from any of its consequences, would be) is likewise something quite incomprehensible to the creator of language and something not in the least worth striving for. This creator only designates the relations of things to men, and for expressing these relations he lays hold of the boldest metaphors. To begin with, a nerve stimulus is transferred into an image: first metaphor. The image, in turn, is imitated in a sound: second metaphor. And each time there is a complete overleaping of one sphere, right into the middle of an entirely new and different one. (TL, 82)

While the neurophysiological research Nietzsche consulted included ample reflections on the significance of its findings for language, it is the resoluteness with which Nietzsche draws the lines between nerve stimulus, mental image, and language that chart philosophically new ground. Nietzsche conceives of each as radically separated from the other, as each constituting a different "sphere," a metaphor that deserves attention. In ancient astronomy, spheres were derived from observations of the movement and the stability in the relations between the stars and planets. The spheres themselves were thought to be invisible, sustaining the phenomenological world without themselves coming into view. With the sphere, Nietzsche introduces ideas of completion and self-containment for each of the receptive areas he identifies, envisioning them as universes of sorts, as observational planes that are closed onto themselves.

As the metaphor emphasizes closure and heterogeneity between nerve stimuli, mental images, and language, it puts more weight on the question of the nature of the relationship or interaction between them. There is quite a bit of semantic wavering regarding this question in the essay. Nietzsche describes it as "a tactile game" (*tastendes Spiel*) and subsequently as a word providing a "representation of the nerve stimulus" (*Abbildung des Nervenreizes*), and then he explores the literal meaning of metaphor in a series of verbs: to transfer (*übertragen*), to fashion after (*nachformen*), and to skip (*überspringen*) (later in the essay, he also uses the verb "to translate" [*übersetzen*]). As Emden has shown, these are all verbs nineteenth-century physiology uses to denote the relation between initial nerve stimulation and subsequent mental states.[24] Yet, all these verbs remain quite ambivalent with regard to the question of whether the relationship between spheres is to be conceived as something that is in some form or another still representational or imitative. Or do we have to think of transference, to use a twentieth-century distinction, as devoid of information, as the mere release of "energy" that would support autoreferentiality in the way implied by the earlier metaphor of tactile play and by Müller's principle of specific nerve energies? It is only in the subsequent example of Ernst Chladni's sound figures,[25] where Müller's principle finds its most ostensible reception, that the relationship between what are perceived as fully heterogeneous spheres receives a conceptual model that aligns rather clearly with the latter position:

> One can imagine a man who is totally deaf and has never had a sensation of sound and music. Perhaps such a person will gaze with astonishment at Chladni's sound figures; perhaps he will discover their causes in the vibrations of the string and will now swear that he must know what men mean by "sound." It is this way with all of us concerning language; we believe that we know something about the things themselves when we speak of trees, colors, snow, and flowers; and yet we possess nothing but metaphors for things—metaphors which correspond in no way to the original entities. In the same way that the sound appears as a sand figure, so the mysterious X of the thing in itself first appears as a nerve stimulus, then as an image, and finally as a sound. Thus the genesis of language does not proceed logically in any case, and all the material within and with which the man of truth, the scientist, and the philosopher later work and build, if not derived from never-never land, is at least not derived from the essence of things. (TL, 82–83)

When Nietzsche juxtaposes visual with auditory perception, he simultaneously exchanges the underlying cognitive model, replacing one that is derived from visual observation with a model that is derived from auditory processes. That is, the relationship between spheres is no longer thought in representational terms along the lines of a subject–object correlation, or as the mind mirroring nature as the visual metaphor would have it, but as a form of resonance. The metaphor thus anticipates Thomas Fuchs's recent description of Varela's and Thompson's notion of enactive embodiment. Fuchs argues that, "instead of a representative or mapping relation, we should rather speak of a continuous mutual resonance between the brain and body."[26] The auditory metaphor implies further that, as with a piano or other musical instrument, the particular tone or quality of the resonance is not a more or less adequate reproduction of an "outside," but the product of the resonating body, of the receptive instrument, and by extension, of the specific sense. That we are dealing with sense-specific outputs is highlighted by Chladni's sand figures. The same stimulus that produces sound for a hearing person may also produce sand figures that can be perceived visually. Such visual representations of sound, however, give the observer no sense of what it means to hear.[27] Hearing hinges on the specific output capabilities of the ear and how the brain renders that output, just as the forms one perceives visually must be thought to reflect the brain's rendering of specific forms or figures the eyes create when stimulated.

We should note that the shapes and images that form on a resonating plate covered with sufficiently moveable elements (such as sand) are not imprints in the sense that classic epistemes from Plato to Locke often figured sensory perception. The figures that form on the resonating plane do not replicate or represent even *ex negativo* any curves or formations that would be properties of the source. This is not to say that an observer cannot construe a correlation between pitch and the image that forms, but rather that information derived from changes in imagery/pitch is the product of the apparatus, not something that would exist independent of the observing organ or mechanism. Perception, Chladni's sand figures suggest, is a reflection of the senses' specific ability to react in sense-specific ways to stimuli, not a replication, representation, or reproduction of qualities of the source. Any information about the stimulus, then, is a secondary act, derived from coordinating and correlating internal differentiations and attributing them to an outside, allowing nevertheless for an amazing amount of coordination between internally derived (i.e., organ-specific) differentiations and their attribution to a physical outside. If it is possible to align Nietzsche's epistemology thus with neocybernetic theories of cognition, it is because they share in Müller's work a common ancestry. They both adopt an organicist model of causation that implies an operational constructivism that Nietzsche extends to language (as the next section will examine more closely) and that stipulates both for consciousness and for language that what they perceive or describe is specific to their operations, existing as observed only on their "insides," as mental images or words.

Nietzsche's Organicist Model of Language

The organicist model Nietzsche adopts from Müller and with which he anticipates twentieth-century neocybernetic theories differs from dominant reductionist and materialist models of cognition in the life sciences crucially in how it defines causality. To begin with, Müller's principle of specific nerve energies suggests a circular process in which the mind (what is perceived or imagined) registers not an outside, but rather reactions by the underlying sensory apparatus (or neural networks) to stimuli. This insight necessitates a change in focus for the cognitive scientist, from considerations of the relationship between inner and outer, thinking and being, subject and object, or in neurological terms, between the source of a stimulus and the neural response to the stimulus, to the relationship between physiology and psychology. At issue now is the role neurophysiological activities play and their relation to the mind or to the imagination—that is, the translation of physiological activities into more or less coherent sensations, mental images, visual representations, and so on. This shift in focus is at the center of Müller's *On Imagined Visual Representations,* a text he published the same year as, and called the continuation of, *Zur vergleichenden Physiologie.* Müller motivates the treatise's task to examine "the visual sense in its interaction with the mind [*Geistesleben*]"[28] first by explaining how, in the organic world, cause and effect must be distinguished from physical and chemical reactions. In a mechanical reaction, a body communicates its quality to another (e.g., speed), and in a chemical reaction two sides combine to form a third that hides (*verschweigt*) the particular qualities of each side (e.g., H_2O). In organic activity (*Wirksamkeit*), however, no quality is communicated, nor do two entities combine to form a third; rather, in organic interactions, the affected side merely reveals an "essential quality" that exists independent of the quality of the source.[29] This particular form of organicist causation is central to the principle of specific nerve energies. It is why Müller can claim, as we saw above with regard to the eye, that it "does not matter what irritates the eye, it may be pushed, pulled, pressed, or galvanized, or it might sense the irritations transmitted sympathetically from other organs; for all these different causes, ... the light nerve will sense its affection as sensation of light, and sense itself as dark when resting."[30] What is being observed is caused by the nerves that are being stimulated and is nerve specific. Light and colors will appear when the visual sensory substance is being stimulated, and darkness in the absence of stimulation.[31]

Nietzsche adopts this organicist model of causation when he describes the relationship between nerve stimuli and mental images as a jumping into completely different spheres. With the reference to Chladni's sand figures, Nietzsche goes one important step further than Müller (and Schopenhauer) before him. He transposes the stated heterogeneity between the different senses, such as between hearing and seeing, also onto the relationship between mental image and language. Just as visually perceived sand figures do not constitute or tell us anything essential about sound, Nietzsche also asks us to think of words that make up language as not constituting or telling us anything essential about our mental images, about what it means to perceive, imagine, or feel. As we saw, moving from one area to another represents a "complete overleaping of one sphere, right into the middle of an entirely new and different one" (TL, 82). When Nietzsche insists on the separateness of these "spheres," he applies the organicist model of the senses to the relationship between consciousness and language. Each are viewed as different sensory "organs," as measuring their own output capabilities without having access to the world of the other or to a supersensory, nonlinguistic reality. With this bold step, Nietzsche ties his philosophy of language to an epistemology that is neither reductionist nor materialist. From his early theories of the metaphoricity of language to his later critique of the seductions of language guiding philosophy and its infatuation with the "subject," Nietzsche's thinking builds on the idea that language creates its own reality rather than reproduces a preexisting order.

Nietzsche expands on his treatment of language as organ and separate sphere in the opening chapter of *Human, All Too Human*, entitled "On First and Last Things." It is here that Nietzsche spells out most clearly how the neurophysiological research he consulted and transposed onto language challenges the all-too-human in traditional conceptions of cognition. In particular, he uses the findings from the life sciences to refute the stability and coherence of the subject as he questions the idea that reason could serve as a stable foundation, as an Archimedean point of sorts that would guarantee the accessibility and reliability of what is being perceived. Again, the reference for Nietzsche is that of the natural sciences. In its opening aphorism, Nietzsche finds chemistry to offer an alternative conceptual model, a model that would guard against a false reliance on an either/or logic that fails to understand how something can derive from its opposite (the erroneous belief in bivalency is also raised prominently in the first aphorism of part 1 of *Beyond Good and Evil,* "On the Prejudices of Philosophers"). Prefiguring his concept of genealogy, Nietzsche ties this observation to the need of a "historical philosophy . . . which can no longer be separated from natural science" (HH I: 12). When Nietzsche speculates that such a use of chemistry must conclude that, in the world of our ideas and sensations, "the most glorious colours are derived from base, indeed from despised materials" (HH I: 12), the invocation of a turn to physiology is more than merely metaphorical. Rather, he treats language itself as a sensory organ that is closed in the very sense suggested by physiology. Nietzsche continues the dovetailing of physiology with language in the next aphorism, identifying "lack of historical sense" as the "family failing [*Erbfehler*] of all philosophers. . . . They will not learn that man has become, that the faculty of cognition [*Erkenntnisvermögen*] has become" (HH I: 13). Nietzsche also transfers the dimension of time from the organic onto cognition. In a strict sense, introducing historicity into cognition is significant only if one assumes that the observing instruments indeed define what is being observed. Only then, changes in the sensory apparatus and in language can be assumed effectively to change what are observable phenomena and how they appear.

Physiology, then, leads Nietzsche to replace the dichotomy between the phenomenal and the noumenal worlds with the distinction between the world as rendered physiologically and the world of language. Once metaphysics will have lost its sway—that is, once the origin of religion, art, and morality can be "perfectly understood without the postulation of metaphysical interference at the commencement or in the course of their progress" (HH I: 16)—Nietzsche sees our interest in the thing in itself and in appearance wane. Instead, an examination of the observing instruments is needed: "The question of how our conception of the world could differ so widely from the disclosed nature of the world will with perfect equanimity be relinquished to the physiology and history of the evolution of organisms and concepts" (HH I: 16). By treating concepts like organisms,[32] Nietzsche does not merely project an organicist model onto language, but treats each as its own observational instrument, as each producing the world in its own image. It is important to note that the argument that the conceptual world, and by extension metaphysical world, differs so widely from the world rendered to us physiologically does not rely on granting the latter privileged access to the world. Rather, noticing how each organ renders the world differently is enough to challenge the idea that either has access to a supersensory reality or otherwise could provide a stable ground for cognition.

Nietzsche continues his treatment of language as an organ in the following aphorism. Refuting its scientific value, he instead asks about its cultural function. It is in this context that Nietzsche reconfigures the metaphor of the Archimedean point:

> *Language as putative science.*—The significance of language for the evolution of culture lies in this, that mankind set up in language a separate world beside the other world, a place it took to be so firmly set that, standing upon it, it could lift the rest of the world off its hinges [*aus den Angeln*] and make itself master of it. To the extent that man has for long ages believed in the concepts and names of things as in *aeternae veritates* he has appropriated to himself that pride by which he raised himself above the animal: he really thought that in language he possessed knowledge of the world. . . . A great deal later—only now—it dawns on men that in their belief in language they have propagated a tremendous error. (HH I: 16)

This passage recalls the opening paragraph of *On Truth and Lies,* where Nietzsche questions the presumptuousness of humanity, which behaves "as though the world's axis [*die Angeln der Welt*] turned within it" (TL, 79).[33] Nietzsche takes the assumption of a pivot or firm point from which to lift the world off its hinges and transcribes it into the mere appearance of lifting, not the world, but us humans above the world. By identifying language as the supposed *terra firma* and "first stage of the occupation with science," the lifting mechanism no longer relies on lever and pivot, however, but inflates the point, turning it into a world upon itself. This point-turned-world does not rest on a firm ground, nor does it provide a firm ground. Language (and, as we read on, with it reason, science, and even logic and mathematics) creates its own, alternative, seemingly free-floating world, a world of its own creation "with which nothing in the real world corresponds" (HH I: 16).[34]

The metaphor shows how Nietzsche identifies an erroneous reliance on language as the basis for humans having separated themselves from and lifted themselves above the world. The basic idea of a reliable correspondence between word and world lies at the heart of humanism's sense of dominance over its surroundings. While Nietzsche affirms a long-standing tradition that suggests that language is what separates humans from other species, he does so not to affirm this humanist conceit, but to argue that it is not language *per se,* but a blindness toward the nature of language and the relationship between language and thing that creates the false sense of superiority and dominance over the world. Nietzsche specifies the character of language, and why it provides an unstable foundation, in the following aphorism, singling out in particular language's ability to make the unequal seem equal and to insinuate temporal stability where there is none. Both acts produce a false sense of identity, which forms the basis for the "evolution of reason" built on the presupposition "that there are identical things, that the same thing is identical at different points of time" (HH I: 16). Logic and mathematics here "would certainly not have come into existence if one had known from the beginning that there was in nature no exactly straight line, no real circle, no absolute magnitude." What for the anthropological discourses of many centuries was considered to be the basis of what lifted humankind above the world of nature, this Nietzsche suggests is built on "imaginary" categories that do not derive from what they construct as their outside. The "sphere" of language is a bubble of sorts, an echo chamber, to use a currently popular term. Its most important feature, the condition of possibility for the creation of a world above and beyond the physiological reality we experience, is its separateness from what our sensory apparatus recognizes as the (sensory-specific) nature of things. Put more positively, Nietzsche views language and consciousness as instruments in the musical sense, as resonating bodies that react to selective stimuli in their instrument-specific ways. In a more contemporary language, that means that he conceives consciousness and language as closed systems, but in a closure that is the precondition for their openness, their ability to react to stimuli and create and then relate to an outside they build on their inside, neurophysiologically or linguistically. If his reasoning is decidedly posthumanist, it is because Nietzsche recognizes how the science of his day challenges not just the reliability, but the presumed exteriority and superiority of the observational capacities of human consciousness and language. Just as astronomy came to undermine the centrality and importance of earth, as Nietzsche alludes to in the opening gambit of *On Truth and Lies,* so neurophysiology has led the intellect to recognize its own limits, undermining one of the main levers humanism uses to lift humans above other species.

Nietzsche's Post-Truth Epistemology: Closing the Cognitive Circle

Nietzsche's epistemology does not deny truth per se, only a certain philosophical conception of truth: the still popular idea that truths express a correspondence between intellect and thing (*adaequatio intellectus et rei*), as Aquinas had defined it. Jettisoning this particular conception of truth does not affect the ability to distinguish lies from truth or lead to an anything-goes relativism. Such accusations miss the rigidity and binding nature of Nietzsche's alternative understanding of truth as convention, as "uniformly valid and binding designations invented for things" (TL, 81). As Nietzsche is well aware here, truths as conventions are a "necessity" for humans to be able to "exist socially," to "make peace and ... banish from this world at least the most flagrant *bellum omni contra omnes.*" The recognition that truths are not anchored in an external reality makes truth not less important, but more important, both socially and politically. As "uniformly valid and binding designations invented for things," the proper use of words enables and secures the social existence of humans.

We will return to the social and political implications of Nietzsche's conception of truth as convention below. Concerning Nietzsche's posthumanist epistemology, it is important to note that, if we stipulate with

Nietzsche that in truth there is no truth in the Aquinian sense, we run into a paradox. The scientific foundations for this claim, the insights the nineteenth century gained into the mechanisms and thus limits of cognition, are themselves subject to these limits. Nietzsche's contemporaries are aware of this paradox. In fact, Nietzsche encountered the paradox in Lange's *History of Materialism*, which, next to Schopenhauer's writings, was one of young Nietzsche's most important philosophical reference works. Applying the implications of the physiological findings to physiology itself, Lange argues that, independent of whether we reduce the phenomena of the sensory world to consciousness (*Vorstellung*) or to the mechanisms of the organs,

> as long as they reveal themselves as products of our organization in the broadest sense of the word, . . . the following series of conclusions will follow:
> (1) The sensory world is a product of our organization.
> (2) Our visible (bodily) organs are, like all other parts of the world of appearance, only images of an unknown object.
> (3) The transcendental foundation of our organization therefore remains unknown to us just as much as the things, which affect them. We always only have the products of both in front of us.[35]

We know that Nietzsche took note of Lange's series of conclusions, since in 1866 he repeated them almost verbatim in a letter to his friend Carl von Gersdorff.[36] Applied to Nietzsche's argument in *Human, All Too Human*, what is most at stake is the distinction between the supposed stability of the conceptual world of language, on the one hand, and what is perceived as the dynamism of the physiologically rendered world, on the other. If this distinction itself falls into the sphere of language—and where else would it fall?—it too must be thought to be in flux (with the state of science). In turn, if we recognize the "outside" rendered to us physiologically as amorphous, as continually in motion, as pure energy, or as "will," this insight too must be recognized as one that is made within the medium of language (otherwise we would have no knowledge of it and no means to communicate the insight). Consequently, the recognition of instability, flux, becoming, and so on in fact also insinuates a false sense of stability. Thus, the transcendental foundation of what Nietzsche stipulates throughout his work as nature, life, will, force, and so on also remains unknown to us.

Nietzsche's level of reflexivity on the limits of the knowledge derived from the hard sciences bares special noting, as it undermines what Vanessa Lemm describes as the "implicit critique of so-called postmodernist readings" especially by analytical commentators who "deny that for Nietzsche there is 'no truth only interpretation' because, according to them, Nietzsche is using natural facts to say something true about morality (for instance, that some moral entities do not exist in nature)."[37] Nietzsche is well aware that the findings of the natural sciences are subject to the same epistemological limitations as those of philosophy or other modes of human inquiry. The recognition of this paradox also matters in light of leading strands of contemporary posthumanism and their propensity to promote notions of flux, of becoming, of relationalism, and other concepts as alternatives to the fixity insinuated by older ontologies or the application of subject–object distinctions. Jane Bennett's "material vitalism," for example (inspired by Gilles Deleuze and Félix Guattari), seeks to examine things, not objects, when they appear as "vivid entities not entirely reducible to the contexts in which (human) subjects set them."[38] Francesca Ferrando's philosophical posthumanism subscribes to an ontology of becoming that strives for a "posthumanist overcoming of any strict dichotomy" and argues that, "to become posthuman, we need to reflect on our location in this material, dynamic, and responsive process, that is, existence."[39] To ensure that the "us" here does not reaffirm another "autonomous agent endowed with transcendental consciousness,"[40] one that earlier in the book Ferrando argued posthumanism ought to overcome, it is crucial, I want to suggest, to reflect also on the specific "embeddedness" that allows the modern observer to reflect on existence in such dynamic terms. This level of reflexivity is precisely what Nietzsche offers when he recognizes the problem that physiological knowledge as much as materialism is subject to the limitation not just of physiology (the inability to access a supersensory world) but also of language, the observational instrument within which the observations on the limits of observation are made. You can note the paradox, but not escape that any such observations will redraw distinctions, and hence create (new) dichotomies and reintroduce stabilities even where it contests that there are none.

In *Human, All Too Human,* Nietzsche addresses the problem in aphorism 13, entitled "Logic of the Dream." Nietzsche describes a peculiar temporal inversion in dreams when the person dreaming adds outside causes to their perceptions. In sleep "our nervous system is continually agitated by a multiplicity of inner events," but the dream is "the *seeking and positing of the causes* of this excitement of the sensibilities, that is to say of the supposed causes" (HH I: 17). The sleeping mind integrates stimulations of the sensory apparatus quite naturally into the dream. Nietzsche here mentions straps tied around one's feet becoming snakes or the ringing of bells or the firing of a cannon being accounted in terms of the dream "so that [the dreamer] *believes* that he experiences the cause of the sound first." However, not only does the integration of the external stimulus translate into a misattribution of causes; the identification of this fact also elucidates how the attribution of causality entails a temporal inversion. The "supposed cause is inferred from the effect and introduced *after* the effect: and all with extraordinary rapidity, so that, as with a conjurer [*Taschenspieler*], a confusion of judgment can here arise and successive events appear as simultaneous events or even with the order of their occurrence reversed" (HH I: 18).[41]

The observation is repeated a few sections later under the heading "fundamental questions of metaphysics." There Nietzsche argues that perception itself evolved slowly from "the purblind mole's eyes," and only "when the various pleasurable and unpleasurable stimuli become more noticeable, various different substances are gradually distinguished" (HH I: 21). Adding evolutionary considerations to Müller's neurophysiology, Nietzsche takes issue with the attribution of causality, reading it as an artificial notion that is added only after the effect: "When the sentient individuum observes itself, it regards every sensation, every change, as something isolated, that is to say unconditioned, disconnected: it emerges out of us independently of anything earlier or later" (HH I: 21). He mentions the feeling of hunger as physically asserting itself "without cause or purpose." With these physiological arguments, Nietzsche challenges the primacy of causality so important still for Schopenhauer. He recognizes it as the (erroneous) mechanism with which our organs produce the semblance of an outside where, to the strict eye of the physiologist, this outside must be recognized as a product of the observer's internal organization.

Nietzsche subsequently links the (false) attribution of causality directly to the structural properties of language. He first returns to mathematics and the problematic assumption of identity, arguing that, since the beginning of times, the laws of numbers were based on the erroneous postulation "that there are identical things," when in fact "nothing is identical with anything else" (HH I: 22). The error produced by the use of arbitrary signs, however, has come to structure our sensations, including our sensations of space and time. The structure of language, in other words, intrudes on sensory experience, which is thought to provide the independent ground on which to base the knowledge of the capabilities and limits of cognition. Language's erroneous "assumptions" about identity and matter, to put it more pointedly, also allow us to recognize the erroneousness of these assumptions. This is precisely the point Nietzsche makes when he argues that a linguistically produced notion of identity forms the basis for science such that:

> The conclusions of science acquire a complete rigorousness and certainty in their coherence with one another; one can build on them—up to that final stage at which our erroneous basic assumptions, those constant errors, come to be incompatible with our conclusions, for example in the theory of atoms. Here we continue to feel ourselves compelled to assume the existence of a "thing" or material "substratum" which is moved, while the whole procedure of science has pursued the task of resolving everything thing-like (material) in motions: here too our sensations divide that which moves from that which is moved, and we cannot get out of this circle because our belief in the existence of things has been tied up with our being from time immemorial. (HH I: 22)[42]

The passage is remarkable not only for how it questions the instruments of scientific inquiry (assumptions of identity, plurality, coherence, time and space), but also for how it situates its own observation within the very parameters of thought it questions. It thus recognizes how the edifice that is science offers also the basis for the deconstruction of science. In its most advanced iteration, science makes discoveries that come to contradict the very foundations on which it is built, dissolving "everything thing-like in motion." The contradiction between matter and energy that the theory of atoms confronted in the late nineteenth century (Nietzsche's source is most likely Lange's discussion of atomism) thereby reflects more broadly the tension between the world of language and what, based on physiology, is recognized as a world dissolved into an amorphous sea of motion. As indicated above, even the latter claim, however, is one made in a medium whose "essence" is to create identities, objects, the semblance of permanence, and so on, and thus cannot claim true access to what it stipulates as its outside.

Nietzsche reflects on the circularity of cognitive processes at crucial moments throughout his work. Early in the essay fragment *On Truth and Lies,* he applies his findings on the limits of perception and language to this finding itself, noting that, once we question a correspondence between word and thing, we also cannot claim with certainty that there is none. An observation such as the opposition "between individual and species is something anthropomorphic and does not originate in the essence of things; although we should not presume to claim that this contrast does not correspond to the essence of things: that would of course be a dogmatic assertion and, as such, would be just as indemonstrable as its opposite" (TL, 83–84). The same limit applies to Nietzsche's literalized use of the concept of metaphor to describe the relationship between nerve stimulus, mental image, and language as transferences. It, too, is twice removed from the "essence of things." It, too, is a metaphor that measures but the output capabilities of linguistic operations.

The paradox remains on Nietzsche's mind. In *Beyond Good and Evil,* written fifteen years after *On Truth and Lies,* Nietzsche raises the problem again, and does so prominently and specifically with reference to physiology:

> To study physiology with a good conscience, we must insist that the sense organs are *not* appearances in the way idealist philosophy uses that term: as such, they certainly could not be causes! Sensualism,

therefore, at least as a regulative principle, if not as a heuristic principle.—What? And other people even say that the external world is the product of our organs? But then our body, as a piece of this external world, would really be the product of our organs! But then our organs themselves would really be—the product of our organs! This looks to me like a thorough *reductio ad absurdum*: given that the concept of a *causa sui* is something thoroughly absurd. So does it follow that the external world is *not* the product of our organs—? (BGE, 15)

The section is intricate, performing itself the recursive operation it describes. While in the essay *On Truth and Lies,* the linguistic problem of being twice removed from a supersensory reality was turned against the statement itself, this time the physiological considerations are turned against physiology. The physiologist has to bracket that our understanding of the sense organs, too, does not represent a physical reality, but merely describes phenomena that themselves are on the order of phenomena. In the language of the younger Nietzsche, they are metaphors, anthropomorphic "truths" that serve a heuristic function, rather than approximate an external world. That is, Nietzsche understands the physiological discipline as providing a conceptual apparatus or "organ" that creates the world in its own image. That the outer world is indeed but a product (*Werk*) of our physiology is precisely what Nietzsche, following Müller,[43] argues in the essay *On Truth and Lies.* As we saw earlier, Nietzsche notes specifically how any inference about the source of nerve stimuli is a misapplication of the principle of sufficient cause. However, once observation is thus turned into auto-observation (this conclusion the younger Nietzsche does not fully spell out), the observer (position) itself must be thought to be the product of the observation (process), whether those observers are thought to be words, nerve cells, or other biological, psychological, or technological "organs." The quoted section in *Beyond Good and Evil,* questioning its own assertions twice, seems on the surface primarily to point toward the problematic status of the concept of causality again, but once observation proceeds along this line, one will necessarily run into the problem of self-causation. At least, that is a necessary consequence once the observer is included as a constitutive part to the observation.

The paradox of recursion plays a fundamental role in Nietzsche's critique of science (see, e.g., GS, 344), of modern culture's continued pursuit of knowledge (see, e.g., GS, 355) and truth, and its reliance on reason. Already in *The Birth of Tragedy,* Nietzsche stages the recognition of the paradox of recursion as a watershed moment in the history of Western civilization. It marks the point where the "optimism" of the Socratic-scientific paradigm has reached its periphery, where "[man] sees to his horror how in these limits logic coils around itself and finally bites its own tail" (BT, 15). From this insight arises the need to replace the drive for truth with a desire for art. David Wellbery reads Nietzsche's use of the snake metaphor, and the paradox of iterative self-application or recursion it marks, as a governing principle of Nietzsche's thought by which he attempts to transgress the limits set by Enlightenment rationality.[44] It is where Nietzsche perhaps breaks most radically with what we saw Claire Colebrook call a "thermodynamic" rationality and a "cosmopolitanism" that fails to change the presumptuousness of the Enlightenment.[45] This break with Enlightenment rationality emerges in Nietzsche's thought as a consequence of his epistemology being infused with findings from nineteenth-century physiology and his reflections on the representational limits of language. Considering that, with the *causa sui* figure, we are dealing with a theological concept that was long reserved for descriptions of God, one might note the irony that the very moment modern physiology hopes to replace metaphysical conceptions of the subject (and with it the distinction between body and mind) with physical and physiological descriptions of cognition, it has to insert (and thus secularize) a highly spiritual concept into its theory. That is only, of course, if we understand the absurdity that is the *causa sui* paradox as a metaphysical concept, rather than as the product and paradoxical precondition for the possibility of observation itself.

Nietzsche never formalizes the paradox of recursion he encounters. As the next chapter will explore in more detail, he instead adopts a sociological conception of truth as convention and explores its critical and political potency. While that sociological conception, mediated by poststructuralism, is central to critical posthumanism, Nietzsche's reflections of the neurophysiological research make him a forerunner of methodological posthumanism, particularly of neocybernetics and second-order systems theory. What else but a *causa sui* paradox is at the center of the operational closure of systems or of the concept of autopoiesis, of systems reproducing the elements that make up the systems that produce them in the first place (e.g., communication producing evermore communication)?[46]

The physiological research of Nietzsche's time discovers the paradox of recursion and eventually confronts it. Paradox here does not imply an error in logic or a contradiction that would disprove the stated thesis. Rather, the paradox *confirms* the thesis, as there is no need to attribute the paradox to the world proper—it is the product of (its) observation (from within). The unavailability of a fixed point, foundation, or otherwise externally anchored truth creates particular challenges for epistemology, but it also presents a form of liberation, the ability to distance, reverse, or otherwise confront existing hierarchies and dogmas. Ironically, then, nineteenth-century science, particularly in the life sciences, is what also frees philosophy from the dictates of science, offering the possibility for a "gay science," a science that would endorse, rather than reject its imaginative dimension.

Viewed from this angle, is Nietzsche's epistemology, which knows no absolute truths, at heart nihilistic? And by extension, is a critical posthumanism that jettisons the foundational role of human consciousness and language also nihilistic? To reject the idea that a constructivist epistemology is inherently nihilistic, we can turn to Niklas Luhmann, one of the most radical methodological posthumanists, whose cybernetically inspired systems theory builds on the neurophysiological tradition associated with Müller's work. In the concluding pages of *Die Wissenschaft der Gesellschaft* (The science of society), Luhmann, in a surprisingly Nietzschean move,[47] compares cognition to art, suggesting that art "serves to make invisible the world as an 'unmarked state' that forms can infringe on, but not represent. Every attempt to do more will have to deal with paradoxical and, respectively, tautological descriptions."[48] Luhmann continues that "a reflection of these circumstances does not have to lead to nihilism, because nihilism makes sense only against the backdrop of an ontological frame of reference that presupposes the distinction between being and not-being."[49] This has led other antifoundationlists to "carry along an 'ultimate symbol' [*Letztsymbol*] such as indescribability, invisibility, or latency," which, in Luhmann's assessment, "merely reflects the contingency of the use of all distinctions. If such self-reflection itself is sustainable, it is because it derives from a form of social differentiation which no longer allows for a binding, authority providing representation of the world within the world, or of society within society."[50] That is, Luhmann closes the circle by attributing the complexity he finds not to the world, but to the structure of society and the observational patterns it exhibits at the end of the twentieth century, including its high degree of reflexivity, its loss of a center indicative of modern society's functional differentiation, and its propensity to include the observer into the observation. The last found its first scientific articulation in nineteenth-century physiology and its first strong philosophical reflection, as I hoped to show, in Nietzsche's work.

Embodiment and Correlationism

I suggested above that Müller's principle of specific nerve energies anticipates what Thompson and Varela describe as "enactive" or "radical-embodiment," a conception of cognition that recognizes how "our environment emerges through embodied interactions that create meaning, and make sense of otherwise arbitrary sensory input."[51] Nietzsche, I argued, adopts and expands Müller's neurophysiological findings, and hence participates in an episteme that has become a key element of what Tamar Sharon characterized as methodological posthumanism. The term "embodiment" itself seems to unite, but also signals divisions within the diverse schools of thoughts associated with posthumanism. N. Katherine Hayles's seminal *How We Became Posthuman* set the course in this regard. Much of her book is about separating posthumanism from cybernetic and information-theoretical constructions of the posthuman in which "embodiment has been systematically downplayed or erased."[52] For Hayles, "information, like humanity, cannot exist apart from the embodiment that brings it into being as a material entity in the world; and embodiment is always instantiated, local, and specific."[53] Hayles uses the term "embodiment" in two complimentary ways. It acts as the moniker for inquiries into physical, physiological, and neurological processes that enable, expand, and limit cognition in humans and animals; and the concept serves to address cultural, technological, and media-technological modes of embeddedness that extend the body's cognitive and precognitive sensibilities, its sense of being ("subjectivity"), and its ethical responsibilities beyond its proper physiological boundaries. Both areas of inquiry play a leading role in posthumanism's dissolution of traditional disciplinary borders, bringing together the humanities with the natural and the social sciences. As Sharon notes in her cartography of posthumanism, the "strong emphasis on materialism (material bodies, physiological processes and more precisely embodiment)"[54] is also what separates radical posthumanism from its poststructuralist and early postmodern heritage.

Understood along these lines, embodiment challenges dualistic modes of thinking that insist on the supremacy and independence of the mind, of consciousness, and of human agency, and thus contributes to what Pramod Nayar describes as posthumanism's predominant concern, the "radical decentering of the traditional sovereign, coherent and autonomous human."[55] Yet, within posthumanism, the discourse of embodiment has also led in different directions. It has invited speculations about *disembodiment* that reinforce rather than challenge notions of sovereignty and autonomy. Fantasies of disembodiment find their most visible expression in pop-culture representations of virtual minds, self-aware cyborgs, computer-generated consciousness, digitally copied souls, and so on. What Hayles calls the "nightmare [of] a culture inhabited by posthumans who regard their bodies as fashion accessories"[56] also fuels transhumanist (Nick Bostrom), metahumanist (Stefan Sorgner), and speculative posthumanist (David Roden) discussions about the ultimate, purportedly disembodied future of the posthuman. As we noted in the previous chapter, these variants of posthumanism are philosophically, epistemologically, and ethically much at odds with critical posthumanism, a tension that is addressed already by Hayles and that is at the center of introductions to posthumanism such as Stefan Herbrechter's

Posthumanism (2013), Cary Wolfe's *What Is Posthumanism?* (2010), and Sharon's chapter "Cartography of Posthumanism" in her book *Human Nature in an Age of Biotechnology* (2014). The main point of contention, as Wolfe argues, is that transhumanists like Bostrom and Hans Moravec continue a fundamental humanist dogma in the idea "that 'the human' is achieved by escaping or repressing not just its animal origins in nature, the biological, and the evolutionary, but more generally by transcending the bonds of materiality and embodiment altogether."[57] Transhumanists and popular futurists who project humanity's future along this line extend a neo-Platonic metaphysical paradigm, a structure we saw Heidegger place at the heart of his extended understanding of humanist thought. For Wolfe and other critical posthumanists, posthumanism is not about conjuring up a biologically posthuman, postembodied, postmaterial reality. It rather represents a *critical* project that builds on the theoretical insights of poststructuralism and neocybernetic theory with the aim of leaving behind the anthropocentric thought patterns and concepts of the humanist tradition, particularly the mind–body dualism as one of its most central instantiations.

Overcoming dualistic modes of thinking is certainly also at the heart of the latest strands of posthumanism, the new materialisms, object-oriented ontology, and speculative realism. These schools tend to radicalize the materialist contention of the discourse on embodiment to the point where the body and notions of embodiment at times fall completely out of view. In their quest to think the "in-itself" not as a correlate between thinking and being or between subject and object, speculative realism, for example, seeks access to a material absolute that it locates beyond and prior to the body and any mode of embodiment. Quentin Meillassoux explicitly spells out this point. He recognizes that the transcendental subject is "indissociable from its incarnation in a body" and grants the body a constitutive role in the "emergence of the conditions for the taking place of the transcendental";[58] but his aim is to go beyond the viewpoint of the transcendental and think the conditions for the appearance of both objective bodies *and* the transcendental subject. The contention of Meillassoux is that it is only in the "ancestral space-time"[59] that lies beyond and prior to the appearance of bodies that we will find the conditions for science, in particular for "a mathematized science ... able to deploy a world that is *separable* from man."[60]

In her latest publications, Hayles has welcomed the new materialisms. She finds that, after the "baroque intricacies of the linguistic turn," these approaches (she mentions in particular Bennett, Karen Barad, Luciana Parisi, Elizabeth Grosz, and Jussi Parikka) "arrive like bursts of oxygen to a fatigued brain [as they] introduce materiality, along with its complex interactions, into humanities discourses that for too long and too often have been oblivious to the fact that all higher consciousness and linguistic acts, no matter how sophisticated and abstract, must in the first instance emerge from underlying material processes."[61] Despite such praise, Hayles is not uncritical of the new materialisms. She laments the absence of considerations of consciousness and cognition and notes "a performative contradiction," how "only beings with higher consciousness can read and understand these arguments."[62] While this suggests that consciousness, understanding, and even reading somehow play a role in our consideration of underlying material processes, the new materialisms' failure to include reflections on the observing medium (senses, consciousness, language, technologies) in their reflections of the material makeup of things appears to remain but a surface problem for Hayles, one of neglected content, of the absence of considerations of the material conditions that enable cognition, consciousness, and/or subjectivity, not one that goes to the heart of the chosen methodology itself. Hayles instead offers her concept of the "cognitive nonconscious" to address the material foundations of cognition, the physiological processes that precede consciousness. In her 2017 book *Unthought,* Hayles returns to the question of embodiment with a materialist bent. She thus extends the question of embodiment that was at the center of her 1999 *How We Became Posthuman* to the questions of materiality, expanding the critical examination of the mind–body dualism to mind–matter relations. No matter how fascinating recent research in this area might be, the question itself is not new. It dates back to the nineteenth century, when early neurophysiologists like Wundt, Helmholtz, and Gustav Fechner pursued research into the material make-up and nonconscious workings of the mind while also contemplating the philosophical implications of their findings.

We will return more extensively to a central nineteenth-century conception of a "cognitive nonconscious" in chapter 4, which will be Wundt's and Nietzsche's understanding of will. For now, let's return to Müller, who offers an early example of how the nineteenth century discovers and reflects on the material underpinnings of consciousness. On the surface, Müller's law of specific nerve energies appears to assert a correlationist perspective. Is Müller not (a speculative realist might argue) merely extending what Meillassoux calls the "Kantian catastrophe"[63] into the realm of physiological science? But as we have seen, a closer examination of Müller's work reveals that the constructive role of physiology is more intricate as it asks us to rethink "realism" itself, including the realism of scientific findings. To begin with, Müller's notion of embodiment suggests a circular process in which the mind (what is perceived or imagined) registers not an outside, but rather reactions by the underlying sensory apparatus (or neural networks) to stimuli. To be clear, Müller does *not* imply that everything is subjective or reducible to human consciousness when he suggests that our perceptions are not defined by or do not measure the outside world, but rather the sense-specific output of its sensory apparatus. Instead, his findings emphasize the constitutive role of the underlying physiology (which has formed over millions of years and not just in humans); and it asks us to conceive as acquired the perceptual as well as the physiological relationship to what either and both have learned to recognize and relate to as their environment. Nietzsche builds on his point when he suggests that language is but a shortcut to an evolutionary process that transformed the arbitrary "relationship of the nerve stimulus to the generated image" into what falsely appears to be a necessary and "strictly causal one" when "the same image has been generated a million times and has been handed down for many generations and finally appears on the same occasion every time for all mankind" (TL, 87).

As I suggested above, Müller's research, which informs central aspects of Nietzsche's thinking, anticipates a constructivist epistemology that is not a variant of Kantian correlationism, nor of what Meillassoux calls "strong correlationism," or the "absolutizing of correlation itself" so that "only the relation between subject and object remains, or some other correlation deemed more fundamental."[64] What the neurophysiological research of Nietzsche's time shows is more radical. It does not suggest the primacy of thinking or a (thinking) subject for objects or being to emerge; nor does it hypostasize the correlation between subject and object, or between thinking and being. Instead, it suggests that what is sensed or perceived, and any relation to an outside that is inferred therefrom, is the product not of a correlation, but of a self-differentiation process, of operations that have learned (on their inside) to distinguish in organism-specific ways between inside and outside. Even the most primitive life forms have this ability (e.g., crabs tend not to pinch themselves). We saw Nietzsche extend this organicist model to language, a move that anticipates twentieth-century operational constructivism, which relates this model to more complex, nonorganic modes of observation that draw on language such as to thinking and communication. The point is that it is not from a *correlation* of preexisting, independent entities that a sense of being or of subjects and objects emerge, but from differentiation processes, from a "one" that has learned to make itself different from itself in order to observe itself as different from its environment. This "one" may well be stipulated to be a material substrate, with the caveat, however, that both the sense of materiality and any concept of materiality are specific to the biological, physiological, and semiotic operations that coevolved in ways allowing them to make such observations. An operational constructivism thus is *not* subsuming observation under the aegis of an autonomous human subject, nor does it resort to a materialist absolute. It stipulates instead that stabilities—what an organism, a mind, or society experiences or observes as reality—are the product of recursive operations of interacting systems that have evolved (and continue to evolve) over long periods of time.

Why does it matter? As mentioned at the beginning of this chapter, the impetus of critical posthumanism in the vein of Wolfe has been to effect political and social change by remaining vigilant toward the normative and discriminatory force of reality proclamations. Some of the new materialisms have drifted toward the other side of this equation. Morton, to name one of their politically more outspoken voices, argues that it is on the "terrain of ontology that many of the urgent ecological battles need to be fought."[65] This might be true in the political arena, where invoking the limits of scientific certainty can be a strategy to avoid modification of political positions and actions (e.g., to protect the pocketbooks of party donors). When adopting the organicist or operational constructivism that connects Müller with Nietzsche and neocybernetic epistemologies from the twentieth and twenty-first centuries, however, it is not necessary to deny the reality of scientific findings, but only their grounding in an absolute outside. If we think of reality as the environment within which life forms (plants, animals, humans, societies) have come to secure themselves, including with the help of science and technology, then the constructivist approach will have as an advantage over ontological and materialist approaches the fact that it can account for and encourage the continued differentiation of observation processes (biological, linguistic, communicational, technological, or scientific). If nothing else, this puts a constructivist approach in a better position to adjust to continued changes in the environment. It also allows the development of new concepts that may help alter perceptions of and relations to the environment in a more strategic manner.

Bruce Clarke's 2020 book on *Gaian Systems* offers an example for how an epistemology that adopts an organicist or operational constructivism can approach ecological and environmental challenges differently from how ontological or materialist approaches do. The neocybernetic theorization of Gaia and its extension to the planetary level, Clarke shows, offers a different understanding of the biosphere, one that eludes the insertion of Gaia "into holistic schemes, in favor of system differentiations that factor the system-environment distinction, the alterity of the inside relative to the outside of the system, into the conceptual equation" (16). Clarke argues that, relative to the human-centered neologism of the Anthropocene, "Gaia is the better concept to confront Western modernity in particular with its others and its unintended effects, including an account of humanity's minor part in Earth's geostory" (256–57).

Furthermore, a constructivist epistemology that accepts recursiveness can distinguish more clearly between physiological and psychological realities while also acknowledging the role of the latter, of fantasy, in casting and recasting over time the understanding of the former and its relationship to its (organ-specific) environment. "Perception is impossible without assumptions,"⁶⁶ as Clarke notes with regard to Gaian thought. A physiologically informed epistemology and a philosophically reflective physiology pay attention to the coconstitutive processes that define what and how various observers, including our bodies, other living beings, different cultures, experience reality. It will thus cherish the important role of the imagination (of metaphor and aesthetic relations, as Nietzsche puts it) including in and for the hard sciences. Müller makes this point in his concluding remarks to *On Imagined Visual Representations,* where he attributes both Goethe's poetic genius and his scientific accomplishments to the force of his plastic imagination.⁶⁷ Emphasizing the role of plastic imagination for discoveries in the natural sciences does not lead us away from things, objects, the thickness of the body, or the realities living beings confront, but has the potential to lead us right to them. In this regard, we might view the life sciences themselves as a particular sensory apparatus whose differentiations increase the variability, and with it the adaptability, to what are perceived and experienced as environmental threats.

Nietzsche's focus on the body, as much as his understanding of art and "metaphor" as creating and extending realities, builds on, I hoped to show in this chapter, this physiological viewpoint. As the next chapter will explore further, Nietzsche extends a biological viewpoint also to the social realm, where he sees realities emerge from the contentious interactions between the members of a group and between the members and their environment. Far from anchoring those actions in the intentions of a subject or the strength of their will, Nietzsche denies the individual members a privileged position outside the whole from where they could fully assess the whole or control its doings. Our understanding of Nietzsche's conception of human sociality will once again be enhanced by including his study of the life sciences. In particular, Nietzsche read with great interest the ethological and entomological research of his time, research that shares with contemporary strands of posthumanist thinking their concern with the strange behavior and curious accomplishments of eusocial animals, of swarms and hives.

3 Insect Sociality

Why do our relatives, the animals, not exhibit any such cultural struggle? We do not know. Very probably some of them—the bees, the ants, the termites—strove for thousands of years before they arrived at the State institutions, the distribution of functions and the restrictions on the individual, for which we admire them today. It is the mark of our present condition that we know from our own feelings that we should not think ourselves happy in any of these animal States or in any of the roles assigned in them to the individual.
—Sigmund Freud, *Civilization and Its Discontents*

Madness is rare in the individual—but with groups, parties, peoples, and ages it is the rule.
—Friedrich Nietzsche, *Beyond Good and Evil*

AFTER REJECTING THE LABEL IN 1999,[1] Bruno Latour resuscitated the name "actor-network theory" (ANT) for his sociological theory in the introduction to his 2005 *Reassembling the Social*. While he still thought that the name was awkward, confusing, and even meaningless, he had to admit that it had established itself as the easiest signpost for his approach. Furthermore, he liked how the acronym references the eusocial insect of the same name, making it "perfectly fit for a blind, myopic, workaholic, trail-sniffing, and collective traveler. An ant writing for other ants, this fits my project very well."[2] There is more to this aside than Latour's tongue-in-cheek identification with the supposed work attitudes of ants (supposed because the myrmecological research shows that ants are not nearly as diligent as their reputation would have us believe).[3] Suggesting that he is but an ant writing for other ants underlines Latour's insistence on sociology (or any other discursive practice for that matter) not offering a detached position, an "outside" from where it could observe and master what it describes in an objective way that would be unaffected by its own network of actors within which it is embedded. Furthermore, ants and other colony-forming insects offer formidable examples for "society [being] the consequence of associations and not their cause."[4] In an anthill, structure, order, the completion of functions, and any accompanying cognitive acts are neither attributable to individual members of the hive nor the product of an abstract whole or of the planning or decrees of an elite class. Rather, they form from seemingly random interactions between its members and between the members and the environment in which they are embedded. In this regard, ants demonstrate what ANT sets out to trace: the aggregate of complex relations between individuals and their material environment that combine to limit and enable action at any moment in any given circumstance.

That creatures who individually possess only minimal cognitive and communication abilities can in aggregate accomplish feats such as farming, the domestication of other species, the building of complex ecosystems, or detailed mappings of their environment infringes on what, in the humanist tradition, was thought to be the exclusive providence of advanced civilizations: conscious thinking, reasoning, planning, and the creation of complex edifices and infrastructures. The surprising intelligence of insects asks us to reconceive intelligence itself: it can no longer be restricted to the unique domain of human thinking, but must be understood instead as a basic property of life—presupposing that life is always already embedded and functioning in a particular environment. The dynamics of self-organizing processes that anthills, beehives, wasp nests, and swarms exhibit have also served as models for programmers of artificial intelligence who, since the 1980s, have looked to find, as Eric Bonabeau, Marco Dorigo, and Guy Theraulaz put it, "an alternative way of designing 'intelligent' systems, in which autonomy, emergence, and distributed functioning replace control, preprogramming, and centralization."[5] The absence of hierarchical command and control structures and of centralized communication channels in swarms also caters to the antiauthoritarian ethos of critical posthumanism.

While insects appear quite regularly in the literature on posthumanism (it is certainly no coincidence that the image of a fly adorns the book cover of Cary Wolfe's *What Is Posthumanism?*), Jussi Parikka so far has offered the only detailed analysis of the confluence between entomology and posthumanism. Parikka focuses primarily on strands of posthumanism associated with the works of Gilles Deleuze and Félix Guattari, including Rosi Braidotti's material feminism, Brian Massumi's work on instinct, and the biophilosophy of Eugene Thacker. Following Deleuze and Guattari, Parikka treats insects as media; that is, he examines them "not according to their innate, morphological essences but as expressions of certain movements, sensations, and interactions with their environments."[6] They form assemblages: "compositions, affects, and passages in a state of becoming and a relationality that is the stuff of experience."[7] Parikka emphasizes how insect behavior contests not just command structures, but more fundamentally the borders between biological systems and their environment. They also contest subject-centered models of cognition and communication. Furthermore, by drawing on Deleuze and Guattari, Parikka's study raises the prospect of a different ontology, one that revolves not around fixed entities and essences, but around fluid borders, dynamic structures, relationalities, multitudes, and assemblages.

The convergences between entomology and posthumanism are neither coincidental nor merely marriages of metaphorical convenience. The challenges insects present to anthropo- and species-centric modes of thinking have long put pressure on entomologists to look for or develop alternative conceptual models to understand the strange behavior of their object of study. Deborah Gordon argues that the "history of our understanding of ant behavior is the history of our changing views of how organizations work."[8] Historically, this has led to entomologists looking at one particular strand of thought we today associate with methodological posthumanism, which is cybernetics, and vice versa, with cybernetics looking at the findings of entomology to model its conceptual apparatus. Niels Werber makes the point that, in the cybernetic adoption of entomological research, ants are used no longer merely in analogy to the social behavior of humans, but under the premise that "societies of whatever beings can be described following the same rules, as myrmecology from

Wheeler to Wilson will find time and again."[9] Charlotte Sleigh examines the extensive connections between the history of entomology and that of cybernetics. Based on works such as William Morton Wheeler's lecture "The Ant-Colony as an Organism" from 1910, which "has retained a certain cult status among thinkers of emergence and superorganismic organization,"[10] and the inclusion of prominent entomologists in the Macy Conferences on cybernetics in the late 1940s and early 1950s, Sleigh even contends: "Ants in their then-favored forms of representation helped to create cybernetic science: that they provided a repository of disciplinary and natural historical metaphor from which it was convenient to draw."[11] Prominent entomologists such as the American sociobiologist Edward O. Wilson had in turn drawn on cybernetics and information theory to understand better the behavioral and communication patterns of insects.[12] But reflections on the convergence between entomology and the behavior of human collectives can be traced back to Nietzsche's time, when, as this chapter will show, they indeed also garnered the German philosopher's attention. Not only did Wheeler borrow the term "superorganism" from Nietzsche's contemporary Herbert Spencer, as Gordon notes,[13] but there is a more immediate source that connects Nietzsche to the entomology of Wheeler, and by extension of Wilson. Their work builds on the research of the Swiss entomologist Auguste Forel, whose *Les Fourmis de la Suisse* from 1874 is an important source for the second revised edition of Alfred Espinas's influential *Des sociétés animals* from 1878, a book that we know Nietzsche read with great interest.

Conceptual Challenges of Nineteenth-Century Entomology

As a resident of Switzerland, Nietzsche lived in what in the nineteenth century was a hotbed of entomology. Sleigh lists Charles Bonnet (1720–1798), Henri de Saussure (1829–1905), François and Pierre Huber (1750–1832 and 1777–1840), Forel (1848–1931), Edouard Bugnion (1845–1939), Félix Santschi (1872–1940), Rudolph Brun (1885–1969), and Heinrich Kütter (1896–1990) as the most prominent Swiss entomologists of the period.[14] Their works had drawn the attention of a broad range of scientists and philosophers with whom Nietzsche was familiar, including Arthur Schopenhauer, Charles Darwin, Albert Lange, Eduard von Hartmann, Wilhelm Wundt, Georg Heinrich Schneider, Spencer, and Espinas. Despite the abundance of entomological research that Nietzsche encountered in his readings, compared to animals with more beastly or herd-like qualities, his references to insects have received very little attention in the secondary literature. In the twenty chapters of the 2004 *Nietzschean Bestiary,* for example, no entry deals with insects in Nietzsche's work, though Alan Schrift's contribution examines Nietzsche's spiders, which are studied by entomologists even though they are arachnids, not insects strictly speaking.[15] That the secondary literature all but ignores references to insects in Nietzsche's writing is perhaps not surprising, as they are few and far between. Though infrequently, they nonetheless appear in important places in Nietzsche's work, offering an alternative creature domain, one that is represented neither by one of Nietzsche's favorite animals of dominance (tigers, lions, birds of prey, and the like) nor by the comfort-seeking herd. What distinguishes the behavior of insect colonies from the herd is the absence of an external organizing authority.[16] Insects offer models for what today we call an emergent phenomenon, a property that "comes to be instantiated in a process or entity that emerges in time."[17] Espinas describes such phenomena in Nietzsche's time with reference to the beehive as "the spontaneous character of the organization."[18] What insect hives make apparent already in the nineteenth century are organizations that are the result of bottom-up processes where interacting elements constitute a whole that exhibits properties that cannot be attributed to the elements independent of their interactions.

While ants and other social insects had caught the attention of philosophers as far back as Aristotle, in the nineteenth-century they were no longer of interest merely as social or political analogues.[19] In the eyes of entomologists, they offered insights into the basic nature of instincts and into both animal and human sociality. Both questions go to the heart of Nietzsche's philosophy. Insect behavior, broadly put, demonstrates the efficacy of instinctual and nonconscious motivators in the absence of higher cognitive faculties, an anthropological presupposition that Nietzsche sustains throughout his writings. Furthermore, because of the discrepancy between the achievements of the collective and the limited cognitive and communicative abilities of the individual members of the collective, insects raise fundamental questions about the relationship between, on the one hand, the wants, desires, and reasoning of individual members of a species and, on the other, the functionality of the whole that forms from their interactions. Nietzsche is, of course, also no stranger to this second matter. In the realm of human interaction, tension between individualistic and social forces constitutes a productive center of Nietzsche's thought. From the early dichotomy between the Apollonian and the pre-individualized Dionysian, to his many critiques of what he describes as "herd mentality," to his extensive reflections of the will and the will to power in his later works, the tension keeps motivating Nietzsche's questioning of the humanist tradition's anthropocentrism and its promotion of individualism.[20]

It is with good reason, then, that we should take a closer look at Nietzsche's references to insects. They bring to the fore the social dimension that underlies nineteenth-century thinking about instinct and will. They also will help us test popular conceptions about Nietzsche's understanding of the will and the will to power that see these concepts as expressions of individual agency, autonomy, and political domination, ignoring the complex scientific and philosophical debates about the links between physiological, psychological, and social processes that Nietzsche engages in his writings. When Jacques Derrida asserts, merely as an aside, that Nietzsche might have been lost in his writings, "much as a spider who finds he is unequal to the web he has spun,"[21] he in fact points toward a problem that is central to Nietzsche's thinking. It concerns the insight that, like spiders, ants, and other eusocial animals, humans also do not have access to an outside position from which to assert agency, autonomy, or domination over the whole. The focus on insects, then, is not only in line with a central insight of Latour's ANT; it also offers another vantage point from which to read Nietzsche, as Vanessa Lemm suggests, through the lens of an "affirmative biopolitics" that "sees in the continuity between human and animal life a source of resistance to the project of dominating and controlling life-processes."[22] Furthermore, examining Nietzsche's entomology adds a posthumanist perspective to the more familiar poststructuralist readings of Nietzsche, and conversely adds historical and philosophical context to contemporary posthumanist concerns with emergent properties and environmental embeddedness.

Social Formations and the Neurophysiological Effects of Language

Insects appear in prominent places already in Nietzsche's early writings. There is the famous comparison of humans, "these clever beasts," to the gnat in the opening paragraph of the essay fragment *On Truth and Lies in a Nonmoral Sense* from 1873. Humans, Nietzsche argues, invented knowledge only to come to understand how "aimless and arbitrary" their intellect is, how their pride is no different from the conceit of a gnat: "[The intellect] is human, and only its possessor and begetter takes it so solemnly—as though the world's axis turned in it. But if we could communicate with the gnat, we would learn that he likewise flies through the air with the same solemnity, that he feels the flying center of the universe within himself" (TL, 79). Stefan Herbrechter cites and discusses this passage at the very beginning of his book on *Critical Posthumanism* as an example of Nietzsche challenging "humanism's anthropocentric ideology."[23] We might add: not only does the comparison manage to diminish the value humans put on the intellect, but by countering anthropocentrism with an entomocentrism, Nietzsche also suggests that gnats, like humans, live in an environment that is specific to them and only for this reason affords them a central role within it (an observation Nietzsche repeats with reference to ants twice in *The Wanderer and His Shadow*[24]).

While the gnat contests the human pride in the intellect, it also offers an example for the kind of self-deception that Nietzsche identifies as revealing the true function of the intellect. As a late and rather feeble organ, the intellect helps humans to compensate for their lack of strength in the animal kingdom. Its main function is dissimulation, first presumably to hunt prey or hide from predators, but foremost to survive socially, "as the individual wants to maintain himself against other individuals" (TL, 81). Here, the intellect finds the means for social survival in language, which sets conventions "to banish from the world at least the most flagrant *bellum omni contra omnes*. This peace treaty that brings in its wake something which appears to be the first step toward acquiring that puzzling truth drive: to wit, that which shall count as 'truth' from now on is established." The concept of "truth" as first invented—truth in the moral sense—is not about access to a supersensory world, but serves as a linguistic metaconcept that decides on the proper or improper use of language, whether the conventions are followed or not.[25] While Nietzsche debunks the traditional (Aquinas's) philosophical concept of truth as an adequate relation between intellect and thing, he here emphasizes the pragmatic use of truths and concepts, their role in allowing humans "to exist socially and with the herd."

But how? The question gains in importance if we consider that, just before Nietzsche articulates his thesis regarding truth and lies in the moral sense (i.e., truth as convention), he used the metaphor of the tiger to insist on the power of the human species's beastly existence over the intellect, reversing not just Enlightenment but also long-standing religious presuppositions:

> And woe to that fatal curiosity which might one day have the power to peer out and down through a crack in the chamber of consciousness and then suspect that man is sustained in the indifference of his ignorance by that which is pitiless, greedy, insatiable, and murderous—as if hanging in dreams on the back of a tiger. (TL, 80)

The metaphor simultaneously suggests the impossibility of penetrating the sphere of the instinctual, its primacy and ferociousness (the tiger) and the helplessness of the intellect, which deceives itself into thinking it is in charge, when in reality, it is merely along for the ride.[26] Hanging on the back of the tiger, the intellect, the image implies, does not even seem to look in the right direction, let alone be in a position to steer the forces that drive the humanimal. Consciousness here is cast as a seemingly immaterial bubble, a blown-up membrane that, if not ready to burst, is at the mercy of its animalistic ground, the tiger.

While the tiger metaphor further decenters the idea of a *sovereign, coherent and autonomous human* able to make rational decisions, it puts increased pressure on the question of how language is able to control these "ferocious beasts" and produce a peace treaty of social order among them. Put more simply, how are we to understand the relationship between our animal drives and instincts, on the one hand, and language as a medium for social ordering processes, on the other? Nietzsche's answer is twofold. He recognizes language itself as a product of instinct (a proposition chapter 4 will revisit in more detail, as it challenges the humanist conceit that language is what separates humans from other animal species); and he attributes neurophysiological properties to language, a dulling of the senses and perception that he suggests has stabilizing effects on the human species's animalistic underbelly. It is with reference to these two questions that Nietzsche turns to insects in the essay, comparing the conceptual edifice built by scientists (their *Baugenie*) with the "spiders' webs" and, twice, with the doings of bees. Nietzsche first claims that "man raises himself far above the bee," only inasmuch as the materials used by humans are human-made, whereas the bee uses wax that it "gathers from nature" (TL, 85), and he adds, invoking the neurological effects of language: "Just as the bee simultaneously constructs cells and fills them with honey, so science works unceasingly on this great columbarium of concepts, the graveyard of perceptions" (TL, 88).

Let's return to the initial question, *how* Nietzsche thinks language is able to allow the humanimal "to exist socially and with the herd" (TL, 81). And more specifically, how can Nietzsche maintain his insistence on social hierarchies in light of the social modalities offered by insects? In *On Truth and Lies,* Nietzsche turns to neurophysiology to answer both of these questions. They are addressed most directly in his famous definition of truth as "a moveable host of metaphors, metonymies, and anthropomorphisms" and his subsequent assertion that "truths are illusions which we have *forgotten* are illusions; they are metaphors that have become worn out and have been drained of sensuous force, coins which have lost their embossing and are considered as metal and no longer as coins" (TL, 84). Nietzsche de-essentializes the difference between truth and metaphor by suggesting that, because of their continued use, we merely forgot about the metaphorical quality of what we subsequently call truths. Words like "metaphor" or "concept" can themselves serve as examples here. We use these concepts without remembering their metaphorical quality, that one speaks of a transfer, the other of a manual process of grabbing or holding something (the German word *Begriff* retains the manual aspect of the

Latin verb *concepere*). Forgetting here is not merely something passive humans would suffer, nor is it about a mere lapse of human attentiveness; rather, forgetting is conceived as an active consequence of the use of language. As a storage medium, language, through its extensive use, also has the opposite effect: it makes humans forget. Nietzsche casts the act of forgetting in neurological terms. Metaphors carry a sensuous force that is "dulled" by long usage. Subsequently, truths and metaphors affect the nervous system differently. Truths drain it of its sensuous force; metaphors stimulate such force. In the following paragraph, he applies the distinction between habitual dulling and sensory force to the relationship between language and perception. Language gives humans the "ability to volatilize perceptual metaphors in a schema [*die anschaulichen Metaphern zu einem Schema zu verflüchtigen*], and thus to dissolve an image into a concept" (TL, 86; KSA, 1:881). Language thus offers the means of controlling the "mass of images that originally streamed from the primal faculty of the human imagination like a fiery liquid [*hitzige Flüssigkeit*]" (TL, 86; KSA, 1:883).

As we saw, the second reference to bees in the essay fragment expands on the neurological effects of language, suggesting that scientific concepts are "the graveyard of perceptions" (TL, 88). While it has negative connotations in the comparison to bees, Nietzsche also recognizes language's effects on perception to be an evolutionary accomplishment. Language is a shortcut that establishes the kind of stability that the sensory apparatus physiologically accomplished only over extended periods of time, "when the same image has been generated millions of times and has been handed down for many generations and finally appears on the same occasion every time for all mankind" (TL, 87). The previous chapter expanded on the neurophysiological research that informs Nietzsche's epistemology here. With regard to the social function of language, let's note how the young Nietzsche recognizes as both a necessity and a detriment the ability to build from the self-produced, almost immaterial material that is language. Language's capacity to tame, petrify, and coagulate imagination's "fiery liquid" is the precondition for the possibility of human beings living a life "with any repose, security, and consistency" (TL, 86). But it also has deadening effects on perception and comes at the expense of what Nietzsche calls "the fundamental human drive" (TL, 88), the drive to create metaphors.

The metaphors Nietzsche uses to describe the metaphoricity of language are not innocent. The "host" (*Heer*) and the coin metaphors—both employed as metaphors of metaphor—are insignia of power, of creating social hierarchies, of conquering people, marking territory, enforcing order, enabling exchange, authorizing value, and so on. In this regard, the dulling effects of language on the sensory apparatus matter less individually than socially and politically. The sociopolitical dimension becomes more apparent when Nietzsche subsequently maps the difference between metaphors and truths and their neurological effects onto two different character types: the man of science and the man of intuition. The former obeys and works within the established conventions called "truths," while the latter creates metaphors that one day might become new truths. Rhetorically, Nietzsche gives preference to the latter, the man of intuition who does not suppress the human drive to create metaphors, but enjoys the pleasures associated with being artistically active. While both are said to desire mastery over life, the man of science does so stoically, "by knowing how to meet the principle needs by means of foresight, prudence, and regularity," whereas the man of intuition is portrayed as an "overjoyed hero" who is blind to misery and who is "counting as real only that life which has been disguised as illusion and beauty" (TL, 90). Nietzsche here even projects these different aesthetic attitudes toward language onto a master–slave dichotomy: the men of science are said to perform "slave services" (*Sklavendienste*) as they adhere to truths and concepts, whereas with the men of intuition, "the intellect has thrown the token of bondage from itself."[27]

The argument is a case in point for why the metaphorical language and overall pathos of Nietzsche's writing are in line with his neurophysiological insights. Rhetorical and poetic means are used rather than systematic rigor to overcome the dulling effects of conceptual language, and not least to avoid catering to an attitude of social, political, and philosophical servitude. The activist ethos of his rhetoric helps explain why young Nietzsche does not offer a more balanced view of the interactions between concepts and metaphors, between science and intuition, while he admits, nevertheless, that both are needed and that both fulfill important functions. Yet, by anchoring social attitudes in differing neurological responses to language, young Nietzsche falls short of developing a conceptual frame that could approximate the formation of social entities, along with attitudes or instincts therein, as an emergent phenomenon exemplified by anthills and beehives. We have to turn to Nietzsche's writings of the 1880s to see him develop more radically posthumanist views of sociality, of what Espinas calls the "spontaneous character of every organization."[28] Mediated through his critical engagement with the writings of Spencer, but also those of Espinas, Schneider, and other social theorists of the late nineteenth century, Nietzsche develops a notion of sociality that no longer anchors social attitudes in differing neurological responses, but tries to understand how interactive processes both derive from and operationalize (to use a more current term) those responses in ways that anticipate contemporary descriptions of social structures emerging from simple interactive processes.

Insect Sociality: Nietzsche Reading Alfred Espinas's *On Animal Societies*

In recent years, a growing body of scholarship has examined connections of Nietzsche's philosophy to the biological, ethological, and protosociological research of his time. Regarding the question of a shared animal sociality, Maria Cristina Fornari focuses extensively on the influence of Spencer on Nietzsche's moral thought,[29] while Robert Holub examines Wilhelm Roux, W. H. Rolph, and Carl von Nägele.[30] A source that is particularly relevant for this chapter (and that has generally received less attention) is Espinas's 1877 *On Animal Societies*. The French thinker's comparative study was published in German translation in 1879. Espinas, who was influenced by Auguste Comte and Spencer, is fascinated with the behavior of insects, particularly ants, and frequently draws on Forel's research to underline the ubiquity of social behavior as much in animals as in humans.[31] Nietzsche extensively marked his personal copy of the German translation of Espinas's book, which he bought in 1882,[32] highlighting passages in the margins with one and sometimes two lines or underlining parts of sentences. While markings might not always express agreement—Nietzsche is also known to write choice words like *Esel* (the German word for "ass," as in the animal not known for its smarts) in the margins when he disagrees—they give us a good sense of what interested him most and how closely he followed the existing research on the topic.

Nietzsche's markings of Espinas's book show first and foremost his interest in descriptions of "unconscious activities of groups that have a very different outcome than what the acting individual expects,"[33] activities that lie at the intersection between biological and sociological phenomena. Espinas raises the topic in reference to the most recent research in a lengthy footnote that he added to the second edition of his book (which is the basis for the German translation Nietzsche read).[34] In the footnote, Espinas discusses volume 6 of Spencer's *Principles of Sociology* from 1876 and volumes 1 and 2 of Albert Schäffle's *Bau und Leben des socialen Körpers* (Formation and life of the social body), from 1875 and 1878, respectively. Espinas challenges the line Spencer draws between animal and human societies that sees in the former a concrete whole but in the latter a discrete whole. Espinas suggests that Hartmann's *Philosophie des Unbewussten* (*Philosophy of the Unconscious*) contests the analogy between biological and social units on the same grounds when he argues that the "beehive fulfills all conditions of an organic unit but that of the cohesion and contiguity of the composing elements."[35] Nietzsche marks the Hartmann reference to the beehive, but then also the counter argument that Espinas discusses in the following paragraph drawing on Schäffle's more recent publication. Schäffle has shown that "the cells in a living body are not always contiguous to each other, but are nevertheless connected with each other through less organized substances that he calls intra-cell substance (blood serum, neuroligin), that assume similar roles as suitable materials do for social life (that paths, railroads, telegraphs, and in general the whole possessions of a people play for the needs of social life)."[36] The comparison elides not just the difference between human and animal sociality, but also that between biological and technological "substances" and their function. How much Espinas challenges the border between physiology and sociality is evident again in the last paragraph of this lengthy footnote, where he summarizes George Henry Lewes's 1877 *The Physical Basis of Mind* as having shown that "not the brain thinks and feels, but the human being." Espinas uses this insight to articulate what today we would call a theory of social embodiment, concluding that Lewes's research "invalidates any opposition between the social and the individual body that is based on the assumption that all parts of one are sensing, where in the other only certain parts display this quality."[37] Lewes and Espinas thus stand at the cusp of understanding emergence as a phenomenon that can be observed across very different planes of biological, social, and material self-organization.[38] What Nietzsche marks with multiple lines in the margins amounts to an explanation of why human sociality is essentially no different from animal and insect sociality: they form a concrete whole that fulfills all conditions of an organic unit, *including* that of "the cohesion and contiguity of the composing elements."[39]

Subsequent markings in Nietzsche's copy of the translation of Espinas's *Des sociétés animals* focus on discussions of the makeup of herds. Nietzsche extensively highlights Espinas's description of marsh snipes, who exert a "kind of domination" over the herd as their calls are being imitated by their peers, which Espinas speculates must express a desire to rule, but also to imitate and follow.[40] Nietzsche also notes Espinas's subsequent descriptions of domestication processes, which, after the initial application of force and intimidation, can create friendships of the highest degree between species[41] and even raise the intelligence of domesticated animals like dogs.[42] It is with regard to relations between different species that Espinas returns to the discussion of insects as the only class of animals besides humans that manages to domesticate other animals. Espinas cites Huber's 1810 *The Natural History of Ants,* which notes that ants "keep aphids as if they were milk cows."[43] In sum, then, Espinas's book offers Nietzsche examples of social formations and accomplishments in the animal world that elide much of the traditional differences used to separate them from human sociality. Furthermore, Espinas shows how collectives form through interactive processes, even in instances of hierarchical expressions of power or domination.

"Swarming Forth from the European Beehive"

How does the entomological research affect Nietzsche's theorizing of human sociality? The most significant appearance of insects in Nietzsche's published writings is found in aphorism 206 from *Daybreak,* which was published in 1881, two years after the German translation of Espinas's *Des sociétés animals.* Titled "The Impossible Class," the aphorism addresses what Holub labels the "social question," recommending that the working class "ought to inaugurate an age of great swarming forth from the European beehive such as has never been seen before, and through this act of free emigration in the grand manner to protest against the machine, against capital, and against the choice now threatening them of being *compelled* to become either the slave of the state or the slave of a party of disruption" (D, 206). In *Human, All Too Human* (1878–1880), under the heading "Modern Restlessness," Nietzsche had described already the "inhabitants of Europe" as "swarming among one another like bees and wasps" (HH I: 285). Likening Europe to a beehive, he suggests in *Daybreak*'s 206 that only by swarming out of Europe could the working class free itself from enslavement, from their lives being "*used up,* as part of a machine and as it were a stopgap to fill a hole in human inventiveness!" (D, 206). Nietzsche rejects the idea that higher wages "could lift from them the *essence* of their miserable condition," recognizing that the life of the industrial worker is one of exploitation, loss of freedom, and "impersonal enslavement." The fight for higher wages does not resolve the systemic problem, that is "the folly of the nations—the folly of wanting above all to produce as much as possible to become as rich as possible," where a great "sum of *inner* value is thrown away in pursuit of this external goal!" The problem for the European working class is twofold. Their "*inner* value" is diminished, as they "no longer know what it is to breathe freely," and "no longer possess the slightest power over themselves," and "all too often grow weary of [themselves] like a drink that has been left too long standing." The sense of alienation is fueled further by newspapers, mass education, and political agitation ("socialistic pied-pipers") that create envy, false hope, and derision for contentment, such as for "the freeheartedness of him without needs."[44]

Nietzsche criticizes Enlightenment ideals of education, progress, and social mobility, and even the efforts to create more humane work conditions on two accounts. First, they establish and help maintain a system that alienates the working class from their doings; that is, they rob workers of the ability to identify with and take ownership of their existence. Secondly, he argues that attempts to ameliorate the situation materially are unable to solve the underlying misery, and rather merely help stabilize and prolong the misery for large swaths of the population. Nietzsche instead demands more radical change, the refusal to accept the situation and the working class's lot therein.[45] He suggests the workers of Europe "henceforth to declare themselves *as a class* a human impossibility." Such a declaration would allow workers, instead of defending, and thus affirming, their enslaved humanity, rather to swarm out and seek new human possibilities, a new humanness and a new humaneness. They would say to themselves, "anything rather than further to endure this indecent servitude, rather than to go on becoming soured and malicious and conspiratorial!" (D, 206).

Without denying Nietzsche's Eurocentrism, which subsequently envisions the European working class "better to go abroad and become *master* in new and savage regions of the world," the mastery is not one that aims at extending an existing political power structure or economic system. Rather, Nietzsche sets as a goal the escape from slavery, to become "above all master over myself; to keep moving from place to place for just as long as any sign of slavery seems to threaten me." Nietzsche is advocating a change in values that seems no longer possible within Europe and its existing institutions, but requires that the collective detach itself from the state and from capital. "Only in distant lands and in the undertakings of swarming trains of colonists will it really become clear how much reason and fairness, how much healthy mistrust, mother Europe has embodied in her sons.... Outside of Europe the virtues of Europe will go with their wanderings with these workers; and that which was at home beginning to degenerate into dangerous ill humour and inclination for crime will, once abroad, acquire a wild beautiful naturalness and be called heroism." He concludes the aphorism arguing that Europe losing one fourth of her inhabitants would be as much of a relief for those who leave as it is for Europe, who, in any case, might compensate for the workforce being "a little depleted" by bringing in "numerous *Chinese:* and they will bring with them the modes of life and thought suitable for industrious ants" (D, 206).[46]

In the nineteenth century, it was not quite as far-fetched to imagine

Europeans leaving the continent in droves to found colonies abroad as it might seem today; such endeavors were supported by state policies and encapsulated by von Bülo's famous call for Germany to look for "its place in the sun." Nietzsche was familiar with such undertakings through his in-law Bernhard Foerster, who in the 1880s tried unsuccessfully to start a German colony in Paraguay.[47] From our contemporary perspective, it is also no longer absurd to think that China would be able to provide an effective (cheap and politically oppressed) replacement for much of Europe's industrial working class. If Nietzsche's suggestion is nevertheless rather peculiar, it is because he understands Europe as a human colony in analogy to a beehive. It suggests that Nietzsche understands Europe as a self-organizing whole, where reforms and improvements would merely extend the existing architecture of the hive that has grown in a direction that, in Nietzsche's assessment, is no longer sustainable. More radical change is needed. Insect colonies offer Nietzsche a different model to envision social change, then, and one not based on reform efforts or a revolutionary teleology, but on a replication process, a swarming out that would allow for the formation of a new hive and a fundamental redoing of the remainders of the original colony (Europe). In this context, it is worth noting the military use of the verb "to swarm" (*schwärmen*), which was a technical term in the army for the "dissolution of a strictly organized platoon into elusive individual movements" where soldiers "are partially free to ramble around."[48] This use of the verb reveals the paradoxical duplicity of the term, expressing the simultaneity of aggregation and dissolution that Nietzsche observes in human and animal collectives. While certain values and instincts would survive the process of swarming out—Nietzsche distinguishing between European and Chinese work ethics—they would acquire new meaning and presumably lead to the creation of a different colony with different structures if embedded in new surroundings.

The practicality of Nietzsche's answer to the social question is certainly more than debatable, and his Eurocentric conceits are politically problematic even if situated in the imperialist context of his time. That Nietzsche has little to say about what these future colonies would look like or how they might work politically and economically, however, is perhaps not just an implicit admission of Nietzsche's limited expertise in political theory and economics, but also a recognition of the unpredictability of evolutionary processes. While Nietzsche never developed a politically actionable theory from the entomological and ethological studies he encountered, and also did not fall into the ugly trappings of eugenic thinking, as did entomologists like Forel or Karl Escherich, still entomology continues to inform his thinking about the formation of collectives and of corresponding individuation processes in the 1880s. Unlike in his earlier writings, however, Nietzsche no longer anchors social structures in differing neurophysiological responses of individual group members. Erasing the border between human and animal sociality, the later Nietzsche no longer attributes a privileged position to the subject, mind, or consciousness in the formation of a whole. This whole, his comparison of the working class with beehives makes clear, is not steered from the outside, but evolves in response to its own movements and its environments. In turn, he seems to recognize how changing the environment, such as geographically, holds the promise of changing human attitudes and values.

The entomological research of his time helped the later Nietzsche to conceptualize sociality in more radical posthumanist terms, as deriving from the interactive dynamics and embeddedness of social processes in which none of the members of a group possess an outside position from which to direct or even fully comprehend the whole. In this regard, Nietzsche anticipates Latour's insight about the ant-like position of any observer of society (including that of modern sociologists). The insect view is also in tune with the posthumanism of Deleuze and Guattari and their use of the metaphor of the rhizome to describe social dynamics. It is no coincidence that, in *A Thousand Plateaus,* Deleuze and Guattari use ants as a model to capture the social cohesion that exists despite (or within) the continued disruption they associate with the rhizome.[49] While entomology can help us explain continuities between posthumanist modes of thinking about social formations in Nietzsche's and in our time, this research tradition stands in stark contrast to the popular image of Nietzsche as a philosopher of hyperindividualism as encapsulated, supposedly, by his concept of the will to power. As the next chapter will show, however, Nietzsche's reflections on instincts, the will, and will to power do not contradict, but build on the entomological research he consulted. They do so in ways that further challenge the anthropocentrism of the humanist tradition and that undermine the emphasis on individualism and willpower this tradition continues to promote to this day.

4 Instinct, Will, and the Will to Power

> DAUGHTER: Daddy, what is an instinct?
> FATHER: An instinct, my dear, is an explanatory principle.
> DAUGHTER: But what does it explain?
> FATHER: Anything—almost anything at all. Anything you want it to explain.
> —Gregory Bateson, "Metalogue: What Is an Instinct?"

IN THE PREVIOUS CHAPTER, the focus was on the peculiar sociality and emergent properties Nietzsche invokes in his references to insects, properties he applies to the cultural decline of Europe and the lot of its working class. This chapter returns to Nietzsche's reference to insects with a different focus: Nietzsche's understanding of instincts, drives, and wills. These concepts are, of course, not unrelated to the question of social emergence. Instincts and drives are usually invoked when species-specific—social in the broader sense—behavior is attributed to the unconscious doings of individual members of a group. Nietzsche was familiar with Kant's definition of instinct along these lines. In his *Lectures on Latin Grammar*, Nietzsche finds the essence of instinct captured in Kant's "wonderous antinomy...that something has a purpose without consciousness. This is the essence of instinct" (KGW, 2/2:188). As an unconscious force, instincts speak to a central aspect of Nietzsche's philosophical anthropology, its insistence on the dominance of nonconscious motivators. Inasmuch as Nietzsche also conceives of willing as an unconscious or preconscious activity, will belongs in the same category as instincts and drives. Nevertheless, in the popular reception of his work, Nietzsche's will and will to power often figure as forces that seem to escape the societal dynamics described in the previous chapter. Put bluntly: Aren't Nietzsche's will and "will to power" concepts by which he advocates ideals of agency and autonomy that would offer an escape from the social pressures symbolized by the herd? If so, do they not reintroduce a strong anthropocentric viewpoint into Nietzsche's thinking, and thus run counter to contemporary posthumanist sensibilities? As this chapter hopes to demonstrate, a closer examination of Nietzsche's concepts of instinct, will, and will to power reveals that they do not adopt an essentially anthropocentric position, but open up additional posthumanist perspectives. Nietzsche's understanding of instinct challenges the border between human and animal, his "will" questions the dichotomy between individual and herd, and, as recent research has shown, his will to power extends a line of vitalist thought that anticipates contemporary materialist contentions in the vein of Gilles Deleuze and Félix Guattari, as well as Jane Bennett. At stake is the critique of what chapter 1 identified as the problematic core of humanism. Nietzsche's contentions and the research that informs them support a posthumanism that aims "to dissipate the onto-theological binaries of life/matter, human/animal, will/determination, and organic/inorganic using arguments and other rhetorical means to induce in human bodies an aesthetic-affective openness to material vitality."[1]

Malleable Instincts

Let's return to Nietzsche's insect references in the essay fragment *On Truth and Lies in a Nonmoral Sense*. The previous chapter discussed the gnat in the essay's opening gambit and Nietzsche's comparisons of the conceptual edifice built by scientists (their *Baugenie*) to the "spiders' webs" and, twice, with the doings of bees. My reading showed how Nietzsche attributed neurological effects to language that he expanded into a theory of social differentiation, countering the "slave services" of the man of science with the aesthetic mastery of life by the man of intuition. But when Nietzsche suggests that, "just as the bee simultaneously constructs cells and fills them with honey, so science works unceasingly on this great columbarium of concepts, the graveyard of perceptions" (TL, 88), he also insinuates that what drives the scientist to contribute to the columbarium of concepts that is language is just as much the product of instinct as are the architectural accomplishments of bees.

When Nietzsche conceives language in analogy to a beehive, he views language as something dynamic and expanding, and as enclosing, housing, and nurturing those who build it. Furthermore, the analogy with the beehive suggests that language is instinctual. The latter is an argument Nietzsche first encountered in Eduard von Hartmann's *Philosophy of the Unconscious,* a book Nietzsche read closely, though also with a lot of skepticism. Hartmann understands the formation of language as being the expression of a mass instinct and as comparable to the interactions of ants that result in the formation of an anthill.[2] Hartmann's comparison connects to a larger nineteenth-century debate on the instinctual origins of language. It finds support, for example, in Gustav Gerber's study *Die Sprache als Kunst* (Language as art) from 1871, which Nietzsche references in his lecture notes. Gerber in turn quotes Wilhelm von Humboldt's 1836 *On the Diversity of Human Language Construction and Its Influence on the Mental Development of the Human Species,* which had argued already that "language first formed following the measure of inspiration and the freedom and strength of cooperating mental forces. They had to emanate from all individuals simultaneously.... Albeit dark and faint, this opens up a vista into a time where for us, individuals lose themselves in the mass of the people."[3] But it is Hartmann's comparison of the creation of language to the mass instinct inherent to beehives and anthills that Nietzsche repeats almost verbatim in his *Lectures on Latin Grammar*:

> It remains only to consider language as the product of instinct, as with bees—the anthill, and so on. Instinct, however, is not the result of conscious deliberation, not merely the consequence of physiological organization, not the result of a mechanism inherent to the brain, not the effect of a mechanism that would come from the outside and in essence would be alien to the spirit, but the most proprietary achievement of the individual or the mass, emanating from its character. (KGW, 2/2:186)

Echoing Hartmann, Nietzsche locates instinct not at the beginning, but at the end of processes that lead to certain behavioral patterns and response options that are neither based on conscious decisions nor mere reflexes. That instincts emanate from the character of the individual or mass suggests that they derive from the semipredictable (but not automated) behavioral patterns with which insects or humans relate to and situate themselves in their environment.[4] What the beehive and anthill is for the insect species that form them, language is for humans: a product of instinct, and the way humans find a home, nourishment, and structure in the world.[5] In Nietzsche's time, science had become the chief architect of this home, working "unceasingly on this great columbarium of concepts" (TL, 88).

Claudia Crawford reads Hartmann as helping Nietzsche "visualize a completely unconscious process of thinking which has a certain logic or inferential quality of its own and which functions as a basis for conscious discursive activity."[6] The concept of "unconscious *thinking*" needs clarification, I believe, since the insinuation of thought, even if preceded by "unconscious," presupposes a kind of rationality and logic at the level of the participating individual that overlooks the crucial point of the reference to the beehive and anthill. For, the examples from the world of insects imply that, whatever the individual ant "thinks" cannot explain or serve as the foundation for the formation of the hive or hill. Analogously, just as individual ants have no understanding of the hill they form and maintain, one cannot presuppose the "thoughts" of individuals to precede the creation of language. Any "logic or inferential quality" one might confer onto instincts exists only in relation to the whole (that is language).[7] Put more simply, the primacy lies with the character of the species (for humans expressed in their language), not the unconscious machinations of the individual.

Christian Bertino, who traced Nietzsche's Hartmann-inspired reflections on the origin of language back to Johann Gottfried Herder's privileging of instinct over consciousness, points out how relating language to instinct challenges a central humanist conceit. If human language derives from the natural language of animals, then the *differentia specifica* that separates humans from animals is not a timeless given, but something that evolved.[8] Furthermore, the role attributed to instincts in the development of language is crucial for the self-assessment of humans as reasonable beings. If instincts are seen as playing a significant role in the origin of language, they unsettle our image of humans as "wholly autonomous, self-determining subjects."[9] A closer examination of Nietzsche's references to ants and bees reveals another dimension of posthumanist thinking. When Nietzsche suggests (with Hartmann) that instinct is neither biologically given nor the product of an external process, he collapses the distinction between inside and outside, between the biological and the social. Instinct, then, must be thought to emerge from the mediation between inner and outer, between individual and individuals, between thinking and communication, between the biological and the social. In this sense, Nietzsche anticipates the posthumanist notion of instinct that Jussi Parikka sees evident in insects, where instincts are "functioning as intensities—prelinguistic modes of intertwining the body with its surroundings (other bodies), the resonance of bodies in continuity and movement."[10] For Nietzsche, such intertwining is mediated through language, not in the poststructuralist sense that would point toward the representational limits of language, but in a pragmatic sense that sees language as a medium through which human collectives self-organize and embed themselves in their environment.

To understand instinct as a liminal concept, as mediating between inner and outer, individual and individuals, thinking and communication, the biological and the social, is to recognize the malleability of instincts. Again, we find Nietzsche anticipating contemporary posthumanist concerns. As Brian Massumi argues in his 2015 essay "The Supernormal Animal," we must think of instincts as adaptive. If instinctual behavior was a mere reflex mechanism, it would be unable to respond to changes in the environment. Massumi instead suggests that we understand instinct as jump-starting an active process, which he then likens to the "performance of an 'improvisation'" (7). As "induced improvisation," Massumi contends, instinct is "formally self-causing" (7). At stake in Massumi's redescription of instinct is the precise relationship between external stimuli and the instinctual response. Rather than assume a reflex, and hence an immediate causal link, Massumi develops an openness-from-closure principle that respects the specificity of the reacting entity while also allowing for variation, surplus responses, and creative adaptations: "The accident-rich environment preys upon the instinctive animal. In answer, animal instinct plays upon the environment—in much the sense a musician plays improvisational variations on a theme" (9). Drawing on Deleuze and Guattari's concept of desire, Massumi continues here: "To do justice to the activity of instinct, it is necessary to respect the autonomy of improvised effect with respect to external causation. Instinctive is spontaneously *effective,* in its affective propulsion. It answers external constraint with creative self-variation, pushing beyond the bounds of common measure." With Hartmann and Nietzsche we might add that language offers our species a formidable tool to answer external constraint with creative self-variation.

Charles Darwin had challenged already the rigidity and supposed automatism of instincts. Referencing the entomological research of his time, Darwin refuted any strict separation between instinct and knowing, suggesting instead that a "little dose, as Pierre Huber expresses it, of judgment or reason, often comes into play, even in animals very low in the scale of nature."[11] For Darwin, "the most wonderful instincts with which we are acquainted, namely, those of the hive-bee and of many ants," offer evidence that instincts "could not possibly have been . . . acquired"[12] through habits turning into an inheritance. What the entomological research shows instead—and what conforms with Darwin's law of natural selection—is the existence of multiple, competing instincts that can be partly lost and reacquired, and that can change in response to changes in the environment.[13]

Understanding instinct as malleable and adaptive helps resolve an apparent contradiction in Nietzsche's use of the term that is noted in the *Historical Dictionary of Philosophy*.[14] On the one hand, Nietzsche affirms instinct over reason, for instance in his celebration of an ancient Greek culture that did not ask about the reasons that motivated actions; on the other hand, he seems to reject instinct when he laments how the Christian herd instinct has prevailed over more noble instincts in Western society. The apparent contradiction disappears once instinct is not conceptualized as genetically preprogrammed or teleologically directed by the purpose or needs of the group (as Kant's definition presupposes). Nietzsche, in line with Darwin and Hartmann, treats instinct instead as something that is adaptive and that develops in relation to the environment to which it responds. Nietzsche does not reduce the environment to biological aspects, but includes language and social and cultural factors, factors that are subject to change. From this ecological vantage point, he can thus argue without contradiction that a particular instinct has become dominant over time, and he does not need to preclude the possibility of old instincts having vanished or remaining latently active, or of new instincts emerging.

Wilhelm Wundt's and Nietzsche's Concept of Will

Nietzsche references instincts throughout his writings without ever developing a concise theory of the instinctual. It is safe to assume that this is at least in part due to his interest in the competing concept of the will, which increasingly drew Nietzsche's attention in the 1880s, the time period when he also developed his signature concept of the will to power. While not identical, both instinct and will refer to unconscious drives or forces, and both return us to the problem central to entomology, the question of the relationship between individual doings and the social aggregate. That question is most apparent when we consider how Nietzsche seems to view the social comfort of the herd as that which is detrimental to expressions of will and the will to power. And yet, a closer examination of Nietzsche's argument in the context of the scientific literature he responds to shows how his concepts of the will and will to power challenge such dichotomies.

That insects remain on Nietzsche's mind as he developed his concepts of will and will to power is evident in a series of unpublished notes from 1883 that reference ants and beehives as they consider social formations, including in a note with the title "Companions and Societies" (KSA, 10:463), and in two outlines for the fourth part of *Thus Spoke Zarathustra,* where Nietzsche planned to include a section on "beehives and the worker" (KSA, 10:527) and on "teaching slaves (beehives) how to tolerate repose [*Ruhe*]. More machines. Reconfiguration of the machine into beauty" (KSA, 10:599). For the question at hand, however, what is most revealing is an unpublished note from spring/summer of 1883 that enlists the social behavior of ants to help explain no less than the emergence of signs and understanding from rudimentary interactive processes and neural responses:

> *To express yourself* [*sich mittheilen*] originally, then, is to *expand your power* [*Gewalt*] *over the other*: at the bottom of this drive lies an old sign language—the sign is the (often painful) *embossing of a will onto another will*
>
> > *to make yourself heard through strokes* (ants).
>
> NB: The *injuries* of the other, too, are the *sign language* of the *stronger*.
>
> In this regard, *understanding* originally is a sensation of pain and an acknowledgement of a foreign power. *To understand quickly and easily* becomes preferable (to receive as few blows as possible)
>
> > the quickest mutual understanding is the *least painful relationship to each other*: this is why we strive for it (
>
> > *negative sympathy*—the original creator of the *herd*). (KSA, 10:298)

The unpolished note (I am reproducing the line breaks of the original) offers an evolutionary theory about the emergence of language and understanding or consciousness from basic neurophysiological responses to social interactions, a theory that cuts across species. Nietzsche thereby envisions understanding as a form of social embodiment: in encounters among themselves, members of a collective learn to remember through an "embossing" of another will onto their own, the sensation of pain that is felt when bodies repeatedly collide. Remembered pain is pain that is detached from the actual event that causes the sensation. It is only through repeated exposure that the event becomes comprehensible: the recognition of pain can detach itself from the actual experience of pain, making the anticipation of the event possible, an *understanding* of pain in the absence of the actual experiencing of pain. Understanding on a most basic neurophysiological level, then, is a form of recollection, the identification and reliving of an absent sensation. That the recalling of absent pain is preferable to experiencing actual pain creates an incentive to recognize and avoid pain before it happens.

Nietzsche cites ants here not metaphorically, but to suggest that we are dealing with a rudimentary process of communication ("self-expression") that derives first from the physical encounters of resistance by the members of a species bumping up against each other. It is from the continued impulses the members of an interacting group receive on a basic physiological level that "understanding" emerges. The contentious interactions between the members of a collective create, we would say today, positive and negative feedback loops, which direct the actions of the members of the collective. The desire to avoid pain leads the collective to *drift* toward the formation of herds. In other words—and counterintuitively—the herd, Nietzsche notes, forms not in the absence of contentious interchanges between the members of the collective, but because of them. Only where there is friction, only where bodies collide, can we expect herds to form.

The secondary literature on Nietzsche has long established how his conception of the will responds to the physiological and psychophysiological research of his time. We know that Nietzsche read Otto Caspari, Wilhelm Roux, Otto Schmitz-Dumont, Johannes Gustav Vogt, and Wilhelm Wundt, paying special attention to how their work challenges the borders between the psychological, the physiological, and the material.[15] For our entomological considerations of the question of the will, Georg Heinrich Schneider's 1880 *Der thierische Wille* (The animal will) and the writings of the physiologist and philosopher Wundt (1832–1920) are most informative and deserving of attention. In fact, Nietzsche's unpublished note from 1883 that invokes ants embossing their wills onto each other stands in close proximity to two references to Schneider's *Der thierische Wille*,[16] which Nietzsche must have read or reread in 1883. Schneider's book discusses Alfred Espinas's *Die thierischen Gesellschaften* (On animal societies) extensively, arguing in the introduction that "the spontaneous character of the organization" must be derived not from an understanding of the whole by the individual members, as Espinas does, but from the rudimentary feelings of pleasure or displeasure the individual insects experience.[17] Schneider points to Wundt's 1863 *Vorlesungen über die Menschen- und Thierseele* (Lectures on the human and animal soul; a book Nietzsche included on his reading list in April/May 1868; see KGW, 1/4:572) as the first and more advanced animal psychology that acknowledges how animal behavior must be derived from base drives.[18] Nietzsche's second note citing Schneider goes to the heart of the matter. Nietzsche quotes and then challenges Schneider's assertion that it is perception (*Vorstellung*) that stimulates the strongest drives leading to action, which makes the human will dependent on what it perceives. Nietzsche counters, using two exclamation marks and an emphatic "say I," that it is the drive (*Trieb*) itself that produces the representation (*Vorstellung*) in the first place. Nietzsche subsequently distinguishes between the force (*Kraft*) of the will and the drive (*Trieb*), the use of this force, concluding that to change morality, one has to focus first on increasing the force, then on its use, the how (see KSA, 10:315).

Nietzsche's assertion that a driving force is directing what is being perceived in the first place is, in fact, not an original insight of his, but precisely one of Wundt's central contentions. Michael Cowan points out that "Nietzsche's insistence on seeing the will not as a separate intellectual category but rather as a force inhabiting *all* organic activity (instincts, drives, passions, etc.) takes up a key component of the model of the will laid out by Wilhelm Wundt in his *Grundzüge der physiologischen Psychologie* (1873/4)."[19] While no personal copy of Wundt's seminal work with Nietzsche's markings is available, we know that Nietzsche had followed the research of Wundt (who was teaching at his alma mater in Leipzig) at least since 1868 (see above) and that Wundt continued to stay on his mind throughout the 1870s and 1880s.[20] In fact, as late as on November 8, 1887, Nietzsche requested in a letter to his publisher, Constantin Georg Neumann, that a copy of his new book *On the Genealogy of Morals* be sent to Professor Dr. Wundt in Leipzig (see 946, in KSB, 8:186–88).

While Wundt is certainly only one of many sources that shape Nietzsche's conception of the will,[21] what makes Wundt particularly interesting is how he expands the will to all organic activity and even to the inorganic. Wundt reverses the hierarchy established by Christian Wolff's psychology of faculties (*Vermögenspsychologie*) that led the eighteenth century to assume that feelings precede drives and drives precede wills.

Wundt cites as evidence for the primacy of what he calls the internal activity of willing (*innere Willensthätigkeit*) its link to apperception.[22] That is, he identifies will as being involved in "unconscious apprehension" (*unbewusste Anschauung*) or "unconscious judgments" (*unbewusste Schlüsse*).[23] In sense physiology, willing is an unconscious cognitive act that directs the attention toward something that enters the field of vision. In a nutshell, the argument is that, because sentient beings need to apperceive—that is, cognitively select something within the perceptual field first—internal willing cannot be separated from sensations and drives, but must precede them. Because of its link to apperception, the (internal) will must be viewed as a "foundational fact" (*fundamentale Thatsache*) and sensations and drives must be derived from a constantly effective activity of willing.[24]

Already in an earlier chapter of his *Grundzüge der physiologischen Psychologie* (*Principles of Physiological Psychology*), Wundt had argued that the inner will is intricately linked to consciousness, self-consciousness, and the sense of an "I."[25] He subsequently contends that the will ultimately appears as the only (!) content of self-consciousness, as the will is central in making mental representations appear as separate from itself. Underlying apperception, inner will creates the appearance of a world that is different from one's own personality.[26] For Wundt, then, will is what structures not just human, but any living being's relationship to its environment: will creates the ability of an organism to experience itself as separate from its environment. In this broad sense, even plants exhibit a will, such as when they turn toward the light of the sun.

The primacy of the inner will does not yet answer the question of how external expressions of the will form or when we can start talking about decisions, conscious pursuits, or choices. Wundt first addresses the question by challenging the traditional presupposition that proceeds from the assumption that will and action are separate. Instead, he suggests, we need to assume that the more primary case (that we can still observe in children and "natural men") is that they are not separate, that apperceptions originally are linked closely to physiological reactions.[27] From such basic, instinctual behavior, Wundt derives the sense of freedom that underlies our understanding of external willpower. He argues that drives that lead the physiology to respond reflexively to stimuli through habitual use create the impression that the success of the reaction is one with the sensation that initiated the reaction. Successful outcome and initial sensation merge and, as the latter gains dominance in the relationship between the two, consciousness (*Bewusstsein*) takes the sensation that accompanied the initial movement to be the driving force. This does not suggest choice yet. Only in mature minds, when multiple wills encounter, counter, and possibly cancel the external action, or when an impulse gains overpowering strength and finally directs the will in its way, can a sense of choice and an awareness of freedom emerge. Only through a conflict of forces that *inhibits* immediate reactions, only through constraint can a space open up between sensation and action, between doer and deed, that allows for the observation of choice, freedom, and the attribution of willpower.[28]

In his 1886 *Beyond Good and Evil*, Nietzsche echoes precisely this part of Wundt's argument, which derives the sense of freedom from constraint, and choice from inhibition (*Hemmung*). Nietzsche claims that the philosophical concept of "the will" gives expression but to a "popular prejudice" (BGE, 19). Analogously here to Wundt's observation, he contends that what is called "freedom of the will" is essentially the affect of superiority with respect to something that must obey: "I am free, 'it' must obey."[29] Any willing simultaneously involves multiple acts of obeying and associated "feelings of compulsion, force, pressure, resistance." It is, to return to the ant analogy, a bottom-up process, the clashing of forces, that defines and directs willing, and it is only after the fact, through a secondary act of attribution, that a sense of volition or even mastery is created.

Nietzsche's psychophysiological considerations challenge the basis for the assumption of human autonomy, self-sufficiency, and agency by questioning the primacy of the "ego" or "I." Returning to the deceptive qualities of the intellect, Nietzsche points out how we are in the "habit of ignoring and deceiving ourselves about this duality [of active and passive forces] by means of the synthetic concept of the 'I.'" The "I," just as much as the will, is but a "popular prejudice" that hides the complexity involved in thinking, feeling, willing, and the like. Nietzsche subsequently compares the presumptuousness of the "I" with a political structure in which the one who is willing takes his feeling of pleasure as the commander, and adds to it the feelings of pleasure from the successful instruments that carry out the task, as well as from the useful "under-wills" or under-souls—our body is, after all, only a society constructed out of many souls—. *L'effet c'est moi*: what happens here is what happens in every well-constructed and happy community: the ruling class identifies itself with the successes of the community. (BGE, 19)

It is important to note here that it is a matter of *misattribution* that creates the feeling of command, when in fact the task was carried out by a multiplicity of competing forces. Maudemarie Clark and David Dudrick, in their analysis of this aphorism, argue accordingly that "Nietzsche presents the 'drama of willing,' then, not as part of how we experience willing, but as an account of what willing is, of how willing is actually constituted."[30] In their conclusion, however, they ignore the theatric quality of this act. Commanding is *not* effective, as Clark and Dudrick suggest, "only if one's drives exist in a 'well-constructed and happy commonwealth,'"[31] but rather the sense of command, autonomy, coherence, and sovereignty arises when an "I" or "ruling class" *illegitimately* takes credit for an outcome that emerged from the interactions of a multiplicity of competing wills.[32] This underlines a point that Günter Abel made in his seminal study of the economy and interpretive force of Nietzsche's wills to power: Nietzsche's understanding of power organizations should not be confused with that of Hobbes's *Leviathan*. Nietzsche is interested in "the process and structure of organization, not in hierarchical orders as established by modern states."[33] Nietzsche, we can add, found in the social behavior of insects important impulses and a model for understanding how rudimentary processes can lead to the formation of structures and higher organizations.

If Nietzsche draws on politics to describe mental processes, he does so not merely as a matter of analogy or metaphorical comparison. Rather, the political metaphor allows him to dovetail psychological and social activities in ways that recognizes their reciprocity—a reciprocity we saw to be as inherent to life forms such as insects as it is to humans. This is underlined in the concluding lines of aphorism 19, where he collapses psychological and social willing under the phenomenon of "life":

> All willing is simply a matter of commanding and obeying, on the groundwork, as I have said, of a society constructed out of many "souls": from which a philosopher should claim the right to understand willing itself within the framework [*Gesichtskreis*] of morality: morality understood as a doctrine of the power relations under which the phenomenon of "life" arises. (BGE, 19)

It is easy to ignore the biological presuppositions that inform Nietzsche's conception of life as that which "arises" from power relations. Viewed against the backdrop of Nietzsche's attention to insects, however, his insinuation is not so much normative as it is descriptive: it does not promote but reveals the process that creates the *semblance* of a "well-constructed and happy commonwealth." It approximates the emergence of different orders from power relations that are to be understood as the immediate, but also quite rudimentary, interactions between multiple wills ("ants"). Put differently, Nietzsche describes a bottom-up process where social order emerges as a secondary product of continuing activities that have neither an outside nor an inside. There is no whole of society where a leading group or imperial "I" directs the doings of its members, a misattribution by those who think they have come out on top; nor is the social order the product of the "intentions" of all or part of its individual members. Rather, Nietzsche expresses a psychophysiological viewpoint that tries to understand the formation of inner and outer "orders" as emergent phenomena, as expressions of an economy of wills underlying the basic interactive nature of "life."

My reading aligns for the most part with that of John Richardson, who details (albeit without much recourse to Nietzsche's sources) the psychophysiological conditions from which Nietzsche derives human agency and self, how Nietzsche thinks that the "reality of 'me' ... is the set of drives and affects that, with their shifting strengths, alliances, and competitions, make me up."[34] Richardson identifies as Nietzsche's goal the creation of a new self that "will work to bring its manifold drives and affects under the command of a single will and project ... in a way that preserves and even enhances the diversity of obeying drives that are gathered into this project."[35] Such a new self cannot simply be posited, but a human can at best work toward conditions (values) that allow for the emergence of a new economy of drives that would approximate such a self. Richardson indeed sees Nietzsche conceive major cultural advances as emergent phenomena, most notably when he suggests the "role of the common is thus to sustain the new level inaugurated by individuals [and that the common] thereby serves as a higher platform from which still higher individuals can then emerge."[36] As I hoped to show, entomology contributed to Nietzsche's thinking about the possibility of fundamental changes in social structure and shared values along these lines.

Understanding what Nietzsche calls "life" as an emergent phenomenon also sheds a different light on aphorism 36 of *Beyond Good and Evil*, which contains Nietzsche's most explicit conception of the will to power. If aphorism 19 breaches the distinction between the social and the physiological, aphorism 36 goes one step further by challenging the distinction between the physiological and the physical world. Again, it is helpful to bring in Wundt as background for Nietzsche's argument. Wundt had already dissolved the dichotomy between psychological and physiological, suggesting that "any movement can be understood as the expression of a drive, as a process [*Vorgang*] that corresponds in its outer appearance with an accompanying sentiment." He goes as far as claiming that we must assume that, "even in the smallest element of substance, the atom, an elementary form of drive must be prefigured."[37] Wundt's claim will seem less obscure if we substitute terms like "energy" or "gravity" for drive. What Wundt is suggesting, then, is that energy is an indispensable aspect for orders to emerge, for negentropy, and hence for the possibility of life to form.

Nietzsche extends Wundt's "monistic perspectivism"[38] in aphorism 36, contending that we "understand the mechanistic world as belonging to the same plane of reality as our affects themselves—, as a primitive form of the world of affect, where everything is contained in a powerful unity before [!] branching off and organizing itself in the organic process." As the concluding lines of aphorism 36 make clear, Nietzsche's concept of the will to power adopts Wundt's expansion of will beyond the sphere of the strictly organic:

> Assuming, finally, that we succeeded in explaining our entire life of drives as the organization and outgrowth of one basic form of will (namely, of the will to power, which is my claim); assuming we could trace all organic functions back to this will to power and find that it even solved the problem of procreation and nutrition (which is a single problem); then we will have earned the right to clearly designate all efficacious force as: will to power. The world seen from inside, the world determined and described with respect to its "intelligible character"—would be just this "will to power" and nothing else.[39]

While on the surface this might seem like an esoteric extension of the psychological onto the physical world—a "psychologization of being itself"[40]—it is an attempt to leave behind dualistic modes of thinking based on the available physiological research of his time. Via its extension to the inorganic world, Nietzsche's will to power responds to and thus anticipates the very concerns of a vitalist materialism of Bennett, "to dissipate the onto-theological binaries of life/matter, human/animal, will/determination, and organic/inorganic."[41] Considering the stereotypical readings of Nietzsche, the most important corrective we can take from reading Nietzsche's will and his will to power against the backdrop of Wundt's psychophysiology is that it shows how these central concepts of Nietzsche's thinking are not attempts to rescue an anthropocentric viewpoint and continue to attribute behavior to individuals, but rather pave the way to account for what today are identified as emergent phenomena. Rather than reiterate mind–body, individual–society, or similar dichotomies, Nietzsche's deconstruction of the will as an umbrella concept and his will to power are attempts to model the emergence of complex edifices from simple operations under which physical, psychological, and social phenomena must be thought to arise in the first place.

The Will Itself ... Is the Herd

Against the backdrop of the contours of a theory of social emergence in Nietzsche, I want to turn to two different legacies of Nietzsche's will in the twentieth century. In his important 2008 book *Cult of the Will*, Cowan examines the cultural and political legacy of late nineteenth century psychophilosophies of will through the first half of the twentieth century and up to today. Cowan identifies Nietzsche not so much as the starting or focal point, but as part of a "much more diffuse network of discourses on the will"[42] that contributed to the posthumous success of Nietzsche's writings. Cowan reads the popularity of the discourses of the will and the "enormous energy expended by turn-of-the-century culture on regaining control over the body"[43] as symptoms of and reactions to the cultural decline diagnosed by so many narratives of modernity (Baudelaire being a rare exception). Willpower, understood as the ability to restrain and control yourself, was promoted, Cowan argues, as it "underpinned a fantasy of regaining control over one's destiny in a world that no longer seemed to allow such control."[44]

The discourses that seek remedy for the perceived social ills of modernity in the "triumph of the will" extend, Cowan shows, from the psychophysiology of the late nineteenth century, to efforts in the First World War to harden soldiers for the battlefield, to the Nazis' staging of such supposed triumphs in mass parades, to today's motivational-speaker industry. Worshipers of the will (including, unfortunately, leading Nazi figures) like to enlist Nietzsche as one of their progenitors. This part of Nietzsche's popular reception ignores, however, the scientific underpinnings of Nietzsche's writings on the will, as well as his later critique of any foundationalist uses of the concept. Nietzsche not only recognized the unity of "the will" and willpower as generalizations and illusions (as such, they might still serve a noble function); his reflections on the will also challenge, as argued above, the basic dichotomy between individuals and society that suggests that we could think of individuals as entities that exist independent of their social and material embeddedness or, vice versa, that we could think of "society" as an independent entity that would compromise the "true nature" of its members.

Nietzsche's understanding of society as an aggregate of competing forces necessitates a reassessment of the idea that his philosophy calls for a strengthening of the will in response to the perceived ills or *décadence* of modernity. As Nietzsche's evolutionary theory of social emergence indicated, it is not the absence of will that promotes the formation of herds, but the colliding of wills at a most rudimentary level. By extension, then, wouldn't the strengthening of individual wills also have to lead to a strengthening of the herd? Or, to spell out the paradox more clearly, wasn't the "cult of the will" that reigned in the twentieth century and that continues to hold its sway to this day prone to strengthen the herd, and hence to weaken its individual members? Nazi Germany offers perhaps the clearest example for these dynamics in which affirmations of will offer no escape from the herd, but help constitute it. Under the banner of a "Triumph of the Will," the fiction of a united strength, the Nazis demanded the complete surrender of individual expressions of will in order to assemble masses of people for the nefarious purposes of their murderous ideology.

In his seminal essay "'Disgregation des Willens'—Nietzsche über Individuum und Individualität" ("'Disgregation of the Will': Nietzsche on the Individual and Individuality"), Werner Hamacher spells out the paradox at the bottom of modern affirmations of individuality and recognizes how Nietzsche's strange expression of a "disgregation of the will" speaks to this very problem.[45] Hamacher addresses the paradox first with regard to modern assertions of individuality. In a nutshell, the concept of individuality that allows us to think and communicate about individuality is, as any concept, not unique and singular, and thus fails to make communicable or even imaginable what is "individual and actual" (TL, 83). Or in the more feverous language of Hamacher's semio-ontology: "The conditions for making present what is individual . . . are the conditions of generalizations, which means: of the de-individualization, rape, and destruction of what is individual and even of the possibility of the individual by a universal law."[46] Individuality then, Hamacher concludes, is always only that which "exceeds its empirical appearance, its social and mental identity and its logical form. Individuality is incalculable excess."[47] Becoming individual, for Nietzsche, might remain the highest goal, as Abel argues, but it is possible only by "overcoming and self-overcoming of the universality structures (e.g. of grammar, sociality, institutionality) specific to the character of the species."[48]

Hamacher sees Nietzsche recognize the tension between affirmations of individuality and the medium that performs this affirmation already in the second *Untimely Meditation,* and again in his later reflections on the will. Hamacher argues that Nietzsche does not conceive of the will as the opposite of the herd, but that he equates the two: "The will itself—this implies the strange expression of its disgregation—is the herd."[49] Pointing out that the root of the term is formed by the Latin word *grex,* which means "herd," Hamacher can argue that disgregation is the condition for the possibility to overcome the will that is constituted by the aggregate (the herd). Only in the disgregation, degenerization, and dissolution of the herd, temporary instances or flashes of will (*Willensmomente*) can materialize beyond their subjugation under a type or *archē*. Accordingly, will can escape the herd only in the form of a "disassociation, excess, and remainder"[50] that is "life" for Nietzsche. Freedom of will in Nietzsche, Hamacher concludes, is not experienced as a form of agency, then, but "in the passivity of a disgregation that, without subject, is subject to the laws neither of the will nor of the concept."[51]

Despite its association with modern *décadence,* then, the "disgregation of the will" is not a purely negative development, but harbors the promise of a new beginning. Nietzsche suggested as much already in aphorism 149 of *The Gay Science,* stating that *décadence* is a sign "that the people is already very heterogenous and is starting to break away from crude herd instincts and the morality of customs [*Sittlichkeit der Sitte*]" (GS, 149). Nietzsche himself points out here how this development is often misunderstood: "a notable hovering condition which one usually disparages as decay in morals and corruption, when in fact it announces the maturation of the egg and the impending breaking of the eggshell."[52] *Décadence,* in other words, is indicative of a weakening of the herd that harbors the possibility of the emergence of new hatchlings. It is a sign, to return to Vanessa Lemm's Foucauldian reading of Nietzsche, of the weakening of a civilization process, "whose aim is the preservation of the group at the cost of the normalization of the individual," a weakening that makes possible the emergence of a culture "that liberates animal life from being the object of political domination and exploitation."[53]

Hamacher's poststructuralist lens brings into focus the semiotic presuppositions that subtend Nietzsche's analysis: how Nietzsche includes the medium of observation, language, as a constitutive component of what is being observed, or in this case, of what cannot be observed because of the generalization process that is necessary for language to work. By ontologizing the semiotic quandaries as "disassociation, excess, and remainder," however, Hamacher runs the risk of replicating the opposition between, on the one hand, a mode of "proper," albeit paradoxical, and hence unthinkable, individuality and, on the other, a society that covers up the disgregation it produces by offering false fictions of coherence, such as the paradoxical fiction of a shared (i.e., unindividual) individuality.[54] For Nietzsche, however, fictions or illusions are not *per se* a problem that should be played off against a falsified or epistemologically elusive reality. The question is what kind and to what effect such illusions are employed, what realities they make possible. The problem with the modern fiction of a "common individuality" is that it subtends the formation of a *congregation* that extends the herd mentality fostered by the Judeo-Christian tradition without addressing and exploiting, as a potential source for a radical new beginning, the disgregating effects of modernity—the sign of the cracking of Europe's eggshell.

Nietzsche's Will

As I have shown, the entomological research of his time allows Nietzsche to think of social orders as self-organizing aggregates rather than along the lines of a dichotomy between individuals and societal forces. The understanding of societal orders as aggregates challenges interpretations of Nietzsche that reduce his social theories to dichotomies between the strong and the weak, between the sovereign individual and the herd, and so on, dichotomies that, as we noted with Derrida before, collapse upon closer inspection, as "Nietzsche himself cannot prevent the most puny weakness being at the same time the most vigorous strength."[55] The more pressing question raised by conceptualizations of society as an aggregate concerns our ability to affect social change. How is change possible in the absence of an outside position or Archimedean point from which to move the world?[56] Short of the suggestion we saw Nietzsche make in *Daybreak* that calls for a swarming out, for the disgregation and regregation of the collective in a different environment, the task of the philosopher would seem to be limited to intervening in society by putting pressure on the medium (language) within which society self-organizes. Nietzsche does so critically and constructively. Critically, he exposes, and thus contributes to, the process of disgregation he associates with modern society by subverting dominant humanist ways of thinking, including their inherent anthropocentrism, the norms they establish, the dualisms they replicate, and so on. Constructively, we can see Nietzsche increasingly in his later years focus on values and expectations that subtend to the interactive quality of social relations. Whatever its exact contours, a new, posthumanist collective should form (in Nietzsche's language) from a relationality that is affirmative rather than reactive, noble rather than resentful, excessive rather than restrictive. In themselves, a critical stance and the propagation of particular values that may guide how humans relate to others and the environment within which they find themselves embedded are certainly insufficient answers to address the most pressing problems contemporary society faces. They offer, however, "impetus to oppose valuations and initiate a revaluation and reversal of 'eternal values'; towards those . . . who in the present tie the knots and gather the force that compels the will of millennia into *new* channels" (BGE, 203). Where these channels will or even should lead remains unclear, inasmuch as they are subject to unpredictable evolutionary processes. Close analyses of Nietzsche's work can, however, continue to support its critical and its constructive thrusts and help invite a diversity of viewpoints over the adherence to norms, foster creative responses over affirmations of existing orders, promote affirmation over resentment, and thus help increase the flexibility of the emerging social, material, and technological networks they subtend. How Nietzsche reflects specifically on the latter, and how his reflections on technology expand his posthumanism, is the subject of the next two chapters.

5 Media Technologies of Hominization

> PREMISES OF THE AGE OF MACHINERY.—The press, the machine, the railway, the telegraph are premises of which no one has yet dared to draw the conclusions that will follow in a thousand years.
> —Friedrich Nietzsche, *The Wanderer and His Shadow*

THE CYBORG is the preeminent symbol of posthumanism. It offers a visual shorthand for the essential role technology plays in the decentering of humans in the twentieth and twenty-first centuries. Posthumanism's "cyborg strand," in which group Cary Wolfe includes Donna Haraway, N. Katherine Hayles, Neil Badmington, Elaine Graham, and Nick Bostrom,[1] has focused extensively on the ways technology pushes and transforms the boundaries that defined "the human" in the humanist tradition. Yet, while the cyborg raises prominently the question of technology and its relation to the organic, it also prejudices how we think about technology. This is particularly evident in popular representations of the cyborg in sci-fi books and movies, as well as in the writings of leading proponents of transhumanism (now "humanity +") like Bostrom and Max More and futurists like Ray Kurzweil or Francis Fukuyama, who like to reference Nietzsche's "overhuman" as an ancestor to their imaginings of a future transhuman or posthuman.[2] They apply a concept of technology that acquired its contemporary meaning only in the early twentieth century[3] and that reduces technology to material applications of science, to instruments, gadgets or other electrically or digitally animated contraptions. The cyborg intimates how such contraptions can replace or enhance parts of the human physiology, limbs and organs, as well as their functions, including the cognitive capabilities of humans. Viewed as a prosthesis and extension of the human physiology, however, technology remains inherently external to the organic. Accordingly, such notions of technology tend to reassert, rather than challenge, a biologically essentialized notion of "the human,"[4] subscribing to a philosophical anthropology that casts humans as deficitary beings.[5] Paradoxically, then, the very moment the cyborg makes the possibility of a merger between technology and the organic appear, the figure of the cyborg asserts their heterogeneity: that technology is essentially external to humans.

This chapter adopts a broader conception of technology, one that includes any artifacts with which humans interact, from sticks and stones to simple tools to complex machinery to more and less advanced communication media technologies. Such a broad conception of technology, which includes the techniques they enable, is historically more appropriate for an examination of Nietzsche's concern with it, an area that has received relatively little attention in the secondary literature.[6] It will also make it possible to establish connections between Nietzsche's thought and more recent strands of posthumanism that speculate less about the future promises or evils of technologies and instead examine the historical couplings and coevolution of humans and technology. The contention is that technologies have always played a constitutive role in the evolution of humans: psychologically, socially, and even physiologically. Among philosophers of technology, Bernard Stiegler expanded this perspective to the very beginnings of humanoid cultural production. Drawing on the research of the French paleontologist André Leroi-Gourhan, Stiegler argued that the prehistorical evidence suggests "that it is the tool, that is, *tekhnê*, that invents the human, not the human who invents the technical."[7] The reference is to the time period between the Zinjanthropian and the Neanthropian, when the use of tools is thought to have promoted evolutionary physiological changes, including the development of an upright skeleton and the growth of the cerebral cortex that allowed for an "exteriozation" of the cortex into matter that made possible the kind of plasticity in perception necessary for reflection and the creation of an "interior" or psyche. In a very material sense, then, the paleontological evidence suggests "that the human invents himself in the technical by inventing the tool—by becoming exteriorized techno-logically."[8]

The concern with how "the technical [is] inventing the human, the human inventing the technical"[9] is also at the heart of the most recent iteration of German, post-Kittlerian, media studies that examines what it calls *Kulturtechniken*: cultural techniques or cultivation technologies. Scholars like Bernhard Siegert, Erhard Schüttpelz, Cornelia Visman, or Sybille Krämer, to mention the school's most prominent representatives, focus on the constitutive role material things, technologies, and techniques play in the cultivation of humans. They broaden the perspective, moving away from traditional electronic or digital media to include simple tools like plows and forks, fences or buildings, as well as the material underpinnings of communication media such as paper, the alphabet, or the printing press and the institutions they enable, such as the postal system. Geoffrey Winthrop-Young describes their research as a "posthuman cultural studies," where the "focus is on short-, middle-, and long-term structuration processes that have taken place throughout history on various sub- and supra-human levels."[10] That is, unlike many strands of British and American cultural studies where culture appears as a secondary marker of human activity—bluntly put, to respect and appreciate aesthetic, social, and political diversity[11]—German scholars of *Kulturtechnologie* see culture not as an addendum or adornment to human nature, but examine instead how technologies, techniques, and practices bring forth and structure materially and symbolically the reality of humans (in both senses of the genitive). Once time is included in such considerations, it becomes apparent that humans and their reality evolve with and along the continued evolution of technologies that they promote.[12] Their relation is not deterministic or causal in any linear sense, but better understood along the lines of an enacted embodiment as outlined in chapter 2, as standing in "a relation of implication or global-to-local encompassing."[13]

A closer examination of Nietzsche and his time's understanding of technology is furthermore important as the question of technology has become a point of contention in Vanessa Lemm's recent discussion of Nietzsche's significance for contemporary posthumanism. In the conclusion chapter of her 2020 book *Homo Natura,* entitled "Posthumanism and Community Life," Lemm does more than simply rebuke transhumanists who erroneously think the overhuman condition is "attained by adding or supplementing technology to the human being" (170). She also argues that the central role technology plays for critical posthumanists like Wolfe and Braidotti, for whom "human nature is rejected in favour of 'the posthuman subject as a composite assemblage of human, non-organic, machinic and other elements'" (171),[14] cannot do justice to Nietzsche's posthumanism. The key issue is that Lemm sees Braidotti and Wolfe employing a concept of technology that views it as prosthetic and supplemental to human beings and that thus separates human beings from nature and treats them as essentially deficient (in the sense employed by Helmuth Plessner's philosophical anthropology). For Nietzsche, Lemm argues, it is the "unknowability of drives that grants the transformative capacity of the human being. Nietzsche's philosophical anthropology starts from what Binswanger calls the 'inner history of life' rather than from the functionality of the human body. Whereas technicity always already seeks to enhance this functionality, it has little to say about the cultural self-transformation that engages the human being through its 'inner history' of the body" (171). Against the backdrop of Lemm's critique, this and the next chapter will argue that Nietzsche and nineteenth-century philosophers of technology did not understand technology as supplemental or prosthetic vis-à-vis nature; rather, they understood the relationship between technology and human nature in dialectical terms, and thus as constitutive to basic processes of "hominization" and "cultivation," terms used here in complementary ways to indicated not an absolute difference, but one in degree. The effects of technologies on cognitive functions, emotions, modes of thinking, and social relations associated with the term "cultivation" also make possible the physiological changes associated with the term "hominization."[15] In this regard, we should also remember the agricultural heritage of the term culture. The cultivation of plants does not preclude the uses of technologies. In fact, in Nietzsche's time, which first used the term *Kulturtechnik,* it referred mainly to the draining and irrigation of fields, processes of environmental engineering.[16]

A Brief History of Technology

The fact that Martin Heidegger, Michel Foucault, Jacques Derrida, and Friedrich Kittler are important precursors to this kind of "posthuman cultural studies" provides a first indication of Nietzsche's relevance for it. Before turning to Nietzsche, though, I want to recall briefly the conceptual history of the term "technology" and the debates it generated in the nineteenth century that form the background to Nietzsche's reflections on technology. In ancient Greek, the term *technê* referred to applied knowledge as opposed to theoretical knowledge, *epistêmê*. Greek "techno-*logia*" referenced the *teaching* of crafts, primarily rhetoric.[17] That is, *technologia* recognizes rhetoric as a craft and technique that can be taught and is the subject of practice. In the eighteenth century, an important expansion of the concept of *techno-logia* took place. Early in the century, the term was still used in reference to the teaching of various crafts (most notably by the German Enlightenment philosopher Christian Wolff). Toward the end of the eighteenth century, however, first attempts were made to distill an abstract concept of "technology" from the craft-specific uses of tools and machines. The publications of two lesser-known Germans show how *Technologie* became an abstract concept no longer tied to the teaching of a particular craft. In 1787, George Friedrich von Lambrecht published a textbook that examines the technologies used in various crafts against the shared horizon of their processing of natural resources.[18] The textbook fell between two publications by Johann Beckmann: *Anleitung zur Technologie oder zur Kenntniß der Handwerke, Fabriken und Manufacturen* (Guidelines for technology, or the understanding of artisanry, factories, and manufacture) from 1777 and, published almost thirty years later in 1806, *Entwurf einer allgemeinen Technologie* (Outline of a general technology).[19] Especially the latter takes a more abstract approach, categorizing a vast array of tools and techniques in various crafts to create a Linnaeus of sorts, a taxonomy of technologies with the hope that such categorizations would enable different crafts to learn from each other through the transfer of tools, techniques, and practices. The aims of Beckmann's taxonomy are not philosophical, but economic and political. In the introduction to his book, the cameralist Beckmann argues that his attempt to systematize existing technologies is a necessary step to stay economically competitive with England. Yet, it is the conceptual impact of his taxonomy that is historically important: it treats technology as an independent category, as something that exists, evolves, and impacts society in terms that are no longer bound to its particular uses in individual crafts.

In the nineteenth century, it was Karl Marx who expanded on the economic and political understanding of technology, reflecting extensively on its social and psychological impacts. Marx was first in viewing technologies as cultural techniques that are fundamentally involved in the hominization process. He articulates his sweeping view of technology directly in an early footnote in chapter 15, "Machinery and Industry," of the first volume of *Capital*:

> Technology discloses man's mode of dealing with Nature, the process of production by which, he sustains his life, and thereby also lays bare the mode of formation of his social relations, and of the mental conceptions that flow from them.[20]

Marx recognizes that technologies impact humans in ways that are much more profound than their economic use-value or their potential to alienate the working class might suggest. By defining the processes of production by which we sustain life, technology also structures social relations, and by extension mental capacities. In a strict sense, then, Marx proposes that technology constitutes the material basis of our thinking, including our thinking about social relations, the means of production, and presumably also technology itself. If technology is indeed at the root of our thinking about ourselves, about technology, and about the relationship between the two, it is impossible to clearly separate technology from human nature. Marx's footnote—which encapsulates his well-traced discourse on humanity's "inorganic body" or "body with organs," as interpreted by Gayatari Spivak[21]—thus displaces and deconstructs the border between nature and technology.[22] In a Marxian sense, then, even poverty and alienation are not problems that emerge when machinery corrupts natural human inclinations, but must instead be thought to be the product of certain (established) technologies of hominization clashing with other technologies.

Marx offers an example of this relationship in his discussion of the distinction between tool and machine in the opening paragraphs of chapter 15, to which he added the footnote quoted above. Marx argues that what distinguishes the machine from a tool is that the former no longer uses humans as its motive power, but "something different from man, as, for instance, an animal, water, wind, and so on."[23] The simple distinction is remarkable, as it suggests that different technologies (tools vs. industrial machines) define differently the agency of humans. Moreover, when Marx subsequently argues that it was not the steam engine that gave rise to the industrial revolution, but "the invention of machines that made a revolution in the form of steam-engines necessary,"[24] he recognizes technology as a self-perpetuating force. That is, the technologies that made modern machinery possible replaced human beings not just as motive powers in the production process, but also as drivers of human history. Neither particular human needs nor the genius of inventors, Marx suggests, but technology itself was the motivating power for the invention of the steam engine, and by extension for the industrialization of Western society. Marx thus anticipates a viewpoint that was developed fully, as Eric Schatzberg argues, only in the 1920s by Charles Beard: "Technology itself [is] the motive force of history."[25] Moreover, the historical frame of Marx's observation on motive power supports Erich Hörl's argument that, "instead of 'Anthropocene' we should say 'Technocene.'"[26]

While Marx's magnum opus is titled *Das Kapital* and not *Die Technologie*, it is clear that neither capitalist accumulation nor phenomena like pauperization, and the social and mental effects they imply, could exist without the technologies that enable machineries and corresponding industries and the circulation of capital. This is not to say that Marx is a technological determinist. For one, Marx thinks of capital itself as a technology in the broader sense, as something that, in its circulation and the exchanges it makes possible, structures social relations and the mental conceptions (such as greed, competitiveness, materialism) that flow from them. Furthermore, Marx stipulates in the so-called "Fragment on Machines" that "machinery does not lose its use value as soon as it ceases to be capital,"[27] but holds the key to reduce "the necessary labour of society to a minimum, which then corresponds to the artistic, scientific, etc.

development of the individuals in the time set free, and with the means created, for all of them."[28] Technology retains the ability to change the means of production once again, and along with them, we must assume, the ability to change social relations and the mental conceptions that flow from them. In other words, while technology enabled and made durable the social relations that make up capitalism, its utility also holds a possible key for ending capitalism, lending capitalism, as Marxists would say, the seeds of its own destruction.

A detailed analysis of the political and economic implications of Marx's thinking of technology are beyond the scope of this book on Nietzsche's posthumanism. They deserve mention, though, for how they offer a conception of technology during Nietzsche's lifetime that does not reduce technology to mere gadgets that might enhance or compromise what are seen as biological or mental givens. Marx points toward an understanding of technology that is broader materially (including anything from simple tools to complex machinery) and ideologically, as technologies affect social relations and the values and mental conceptions that flow from them. This concerns the interactions and border between technology and biology as much as how we think of human agency (for example, as a replaceable motive power), human worth ("the production of too many useful things results in too many useless people"),[29] and human relations ("all that is solid melts into air").[30]

Before turning to Nietzsche's reflections on technology, which show some surprising parallels to Marx's, it should be noted that Marx's footnote was written in the early 1860s, anticipating by about a decade the publication of the first philosophy of technology by name, Ernst Kapp's 1872 *Elements of a Philosophy of Technology*. The recently published translation of Kapp's book into English (as well as the excellent introduction by the editors, Jeffrey Kirkwood and Leif Weatherby) has drawn renewed attention to Kapp's work, which has been often mentioned, but rarely discussed in any detail by twentieth-century philosophers of technology, including Kittler, Stiegler, and Don Ihde. Kapp's concept of "organ projection" offers what should be considered the first full-fledged theory of "hominization" in the vein of Stiegler's use of this term. Kapp's philosophy links technology directly to the evolution of the human species. That is, he frames the relationship between human and technological development in dialectical terms. Humans evolve, Kapp suggests, physiologically and cognitively in and through their encounter with material things and vice versa: technology evolves along the evolution of humans that it enables. Kapp puts this dialectic at the very beginning of "primitive man's" cultivation. The early use of sticks and stones, the simplest of tools, did not merely extend the function of the hand or arm; rather, it made possible the evolution of the hand itself, the hand's "gradual acquisition of a degree of tenderness and flexibility since the tool both exercised and protected the hand.... This way, by reciprocal action, the tool supported the evolution of the natural organ, the organ in turn supporting, at each correspondingly higher stage of dexterity, the evolution and perfection of the tool."[31]

Kapp suggests a similar dialectical process produced consciousness and a sense of self that is separate from that of a baboon. It is the ability to "concern oneself" (*sich befassen*) with an object, rather than merely grab and throw it, that produces, as the reflexive pronoun indicates, a sense both of the object *and* of the self that is different for humans than for apes.[32] The German verb *sich befassen* is telling here. Without the reflexive pronoun, it merely means to grab or hold; but with the reflexive pronoun, the manual activity turns into something intellectual, a verb for studying, for a way of concerning yourself with an object in which the object appears in relation to the self that examines it from multiple angles. What brings the two together is the aspect of retention, the safekeeping that is implied by both the grabbing and the holding-on-to the object that is part of you concerning yourself with it. Kapp indeed identifies retention as a central aspect of the hominization process. The example of the stone suggests that such retention is not a mere mental ability, but dependent on the material thing, the stone itself. The stone serves as an external memory that makes remembering possible in the first place. And it is the lingering concern with the exterior stone as a storage medium that produces a sense of (inner) self. Kapp thus already recognizes in 1872 how *homo sapiens sapiens* is, as Helmuth Plessner proclaimed, "artificial by nature,"[33] or as Stiegler put it over a century later, how "the technical [is] inventing the human, the human inventing the technical."[34]

Kapp's *Elements of a Philosophy of Technology* was published only two years before Nietzsche's second *Untimely Meditation*, the text that contains Nietzsche's most direct and extensive discussion of the effects of technology. That only few studies have ventured to examine Nietzsche's writings on technology is largely due to the dominance of the "gadget" understanding of technology I sketched out earlier. However, once we draw on a broader understanding of technology, one more appropriate for the nineteenth century that includes material objects, tools, machines, and the techniques they enable, we will find that technology plays a more central role in Nietzsche's writing than generally granted. While Nietzsche's knowledge of Marx was at best cursory, and while there is also no evidence that he had read Kapp's *Elements of a Philosophy of Technology*, Nietzsche's writings on technology reflect his time's broader interest in the role technology plays in the history of Western civilization—a history that Nietzsche views as a cultivation process that, on closer inspection, turns out to be enabled and structured in crucial ways by technology.

A posthumanist focus on technology as sketched above makes it possible to bring together two seemingly disparate aspects of Nietzsche's writing: his reflections on the culturally ruinous effects of media technologies in the nineteenth century (which anticipate ideas that are fully spelled out only by the fathers of modern media theory, Eric A. Havelock, Walter J. Ong, and Marshall McLuhan), as explicated in his second *Untimely Meditation*, and his most extensive elaboration of the cultivation process that produced us "modern" humans laid out in *On the Genealogy of Morals* (and to which the next chapter will turn in more detail). In both texts, Nietzsche links particular technologies and techniques to the cultivation of humans, to their emotional, intellectual, and social development. One of the central tenets of Nietzsche's philosophy is that the face of this human, to speak with Foucault, is temporary. As the history of Western civilization supposedly reached a turning point with the production of the "ultimate human" (*der letzte Mensch*), the counterfigure to the overhuman, it now ought to struggle to define a new goal, a new purpose, and cultivate a new, posthuman human.[35] In this regard, *Zarathustra*'s concept of the overhuman, Nietzsche's explorations of "free spirit" in *Beyond Good and Evil*, and the "sovereign individual" in *On the Genealogy of Morals* should themselves be seen not merely as prophecies, ideals, or goals, but as experimental techniques with which Nietzsche hopes to affect the ongoing cultivation and hominization process.

Nietzsche's Typewriter and Other Writing Technologies

Kittler took Nietzsche's brief use of a typewriter and subsequent admission in a letter that "our writing materials contribute their part to our thinking"[36] to express a fundamental change in discourse networks. While dominant reading and writing practices around 1800 produced the belief in the magic of letters at the expense of the materiality of the signifier, the era around 1900 was marked by a "histrionics of media"[37] that attested, as David Wellbery states in his foreword to Kittler's *Discourse Networks 1800/1900*, to "the technological dissolution of the noumatic world."[38] Kittler credits Nietzsche with being the first philosopher who recognized the change in paradigms. Attention to the materiality of communication media leads Nietzsche to challenge the myth of a "continuous transition from nature to culture."[39] Technology forges both what is viewed and experienced as culture and what is viewed and experienced as nature or as natural. The constitutive role of technology, Kittler suggests, is evident in its fashioning of the subject. As Kittler puts it, for Nietzsche, writing, "rather than presenting the subject with something to be deciphered, makes the subject what it is."[40] In this regard, Nietzsche's typewriter is a machine that reveals how language is a technology, an exteriority that produces the interiority it purports to express.

With its focus on changes in discourse networks, Kittler's short chapter on Nietzsche's typewriter does not pursue the full extent to which Nietzsche looks at technology as a means of hominization and cultivation. We should remember that Nietzsche lived in a time period that witnessed unprecedented scientific discoveries and technological advances when, as Gregory Moore details with regard to the period between 1820 and 1900, "an overwhelmingly rapid succession of innovations transformed the world and the way people understood it: in engineering, the development of mechanized industry, the railways, telegraphy, synthetic dyes, the telephone, gas and electric lighting; in physics, the articulation of thermodynamics and the rise of statistical mechanics; in the life sciences, the discovery of cell, evolutionary theory, and experimental physiology."[41] How humans evolved along *and* apart from other animals is a crucial question for any philosophical anthropology that accepts that the species *homo sapiens sapiens,* this "clever beast," shares ancestry with apes and is a mammal that evolved along with the rest of the biosphere. Nietzsche's reflections on technology are focused on precisely this question: how technologies, techniques, and practices cultivated our species in ways that produced the modern human with its particular sensibilities, beliefs, values, desires, and so on. What happened as this cultivation process seemed to enter a new frontier in the late nineteenth century, a century of revolutionary scientific and technological development?

The text in which Nietzsche addresses this question most directly with reference to modern technologies is his second *Untimely Meditation,* "On the Uses and Disadvantages of History for Life." Its opening gambit, which presents the reader with a short fable, makes clear that the treatise is concerned not just with history, but with the fundamental anthropological question concerning the difference between human beings and animals:

> Consider the cattle, grazing as they pass you by: they do not know what is meant by yesterday or today, they leap about, eat, rest, digest, leap about again, and so from morn till night and from day to day, fettered to the moment and its pleasure or displeasure, and thus neither melancholy nor bored. This is a hard sight for man to see; for, though he thinks himself better than the animals because he is human, he cannot help envying them for their happiness—what they have, a life neither bored nor painful, is precisely what he wants, yet he cannot have it because he refuses to be like an animal. A human being may well ask an animal: "Why do you not speak to me of your happiness but only stand and gaze at me?" The animal would like to answer, and say, "The reason is I always forget what I was going to say"—but then he forgot this answer too, and stayed silent: so that the human being was left wondering. (HL, 60–61)

The short fable simultaneously asserts and disrupts the hierarchy between humans and animals. While humans *think* themselves better than the animal, they envy the animal for its happiness, a happiness that owes itself to the ability to live in the moment. The privilege of thought is further undermined at the end of the parable when the narrator allows the animal to speak only to leave animal and human in a shared state of wondrous silence. Human thought not only prevents happiness; it also offers no answer to the question of happiness, leaving it wanting and "wondering."

The fable identifies memory as the main differentiator between humans and animals. What makes humans aware of the present, but also prevents them from living *in* the present, is the ability to recollect and make past moments present again, their ability to represent. The following paragraph suggests that representation is what defines the structure of human consciousness.[42] In an incisive metaphor, Nietzsche links the temporal structure of consciousness—its fractured relationship to the present moment—to the material substrate of writing:

> But he also wonders at himself, that he cannot learn to forget but clings relentlessly to the past: however far and fast he may run, this chain runs with him. And it is a matter for wonder: a moment, now here and then gone, nothing before it came, again nothing after it has gone, nonetheless returns as a ghost and disturbs the peace of a later moment. A leaf flutters from the scroll of time, floats away—and suddenly floats back again and falls into the man's lap. Then the man says 'I remember' and envies the animal, who at once forgets and for whom every moment really dies, sinks back into night and fog and is extinguished forever. (HL, 61)

In this remarkably poetic and philosophically profound scene, remembering is not cast as an act or accomplishment of human subjectivity, but as a curse, an involuntary activity that separates human consciousness from the present, and with it from the happy innocence of the animal. Because it is involuntary, the act of representation that defines consciousness is not inherent to consciousness, is not itself the product of a conscious act. Rather, with the metaphor of the leaf/page fluttering from the scroll of time, Nietzsche links the world of representation to the material substrate of writing, the storage medium of one of the oldest communication-media technologies. With the scroll and the leaf, the German word for which (*Blatt*) is also the common term for the page of a book, as well as for a newspaper, Nietzsche links the temporal structure of consciousness, and by extension what separates human beings from the animal, not to language *per se*, but to material artifacts that allow humans to record, store, and recall.

Media-Technological Barbarism

Taken on its own, Nietzsche's metaphor of the page that flutters from the scroll of time to disturb the peace of the moment might be seen as slim evidence for Nietzsche's recognition of the constitutive force of communication media technologies in the hominization process. The metaphor, however, is not the only instance in the second *Untimely Meditation* where Nietzsche's attention would turn to media technologies. Rather, it foreshadows a central argument of the critique of modernity that is at the heart of the essay: Nietzsche's claim that modern society lacks culture altogether. Nietzsche links modernity's loss of culture to a fundamental change in how information is being processed in the aftermath of the Enlightenment. He singles out two technologies as main culprits, the education system and the mass media of his time, newspapers. Their increased social impact in part can be linked to refinements in the printing process that affected not just the availability of newspapers but also publication expectations for professors, turning "men of learning" into "exhausted hens . . . who can only cackle more than ever because they lay eggs more often: though the eggs, to be sure, have got smaller and smaller (though the books have got thicker and thicker)" (HL, 99).[43]

What does the loss of culture Nietzsche laments mean and how is it linked to education and the press? Following Jakob Burckhardt's understanding of culture as an "organic, collective work of art,"[44] Nietzsche defines culture as "the unity of artistic style in all of the expressions of the life of a people" (HL, 79),[45] a definition he returns to at the end of the essay:

> that culture can be something other than a *decoration of life*, that is to say at bottom no more than dissimulation and disguise; for all adornment conceals that which is adorned. Thus the Greek conception of culture will be unveiled to him—in antithesis to the Roman—the conception of culture as a new and improved *physis*, without inner

and outer, without dissimulation and convention, culture as a unanimity of life, thought, appearance and will. (HL, 123)

The passage is revealing, as it does not oppose human nature to culture, but sees nature as a product of culture. The claim is that Greek culture produced a different *physis* than Roman or nineteenth-century European or German culture. Rather than genes, the practices of peoples and ages determine human nature, with "nature" defined here as how life, thought, appearing, and willing coalesce. Following Nietzsche's claim, changes in "human nature" are not the exception, but rather accompany any changes in culture. Nietzsche states this point more explicitly in his discussion of the advantages and disadvantages of critical history. He identifies critical history as a technique that can reverse what is viewed and experienced as natural or contrived. Critical history's battles against settled history show how every "first nature was once a second nature and that every victorious second nature will become a first" (HL, 77).[46] Nature—what is perceived and experienced as natural—is how culture manifests itself. What, then, causes modernity's loss of culture that, following Nietzsche's logic, must also amount to a loss of nature (or instincts, as we saw Nietzsche argue in the previous chapter), a loss of cohesion between life, thought, appearing, and willing?

Kittler attributed the loss of meaning Nietzsche laments to nineteenth-century efforts at mass education. Increasing literacy rates in nineteenth-century Europe (and not just the social media of today) led to a deterioration of reading practices, the skimming of texts.[47] A closer look reveals that the cultivating effects of the modern "educational operation" (HL, 118) and of the mass media are still more pervasive than superficial reading practices might suggest. Nietzsche turns more broadly against the Enlightenment belief in the efficacy of education or *Bildung* understood in the tradition of Johann Wolfgang von Goethe and Alexander von Humboldt as the endeavor to fashion (literarily, to form after an image) a cultured, mature, autonomous, responsible, civilized, and "whole" human being. Rather than provide humans with culture, Nietzsche blames the particular technologies and techniques of knowledge distribution of his time for having the opposite effect:

> The uniform canon is that the young man has to start with a knowledge of culture.... And this knowledge of culture is instilled into the youth in the form of historical knowledge, that is to say, his head is crammed with a tremendous number of ideas derived from a highly indirect knowledge of past ages and peoples, not from direct observation of life. (HL, 118)

Nietzsche describes what is part of the transition from premodern to modern societies. Whereas in the former, culture is acquired through emulation, as part of ordinary socialization processes, the latter has made culture an object of study, and thus has separated it from life. This separation has made culture traits appear as contingent and contrived, including the traits of a culture that makes culture an object of study. Dirk Baecker argues that the term "culture" as it has been used since the late eighteenth century introduces with the sensibilities and respect for other cultures also the knowledge of the contingency of cultural habits, that cultural habits are not natural or God-given, but vary over space and time.[48] The concept itself thus contributes to the creation of the rift opening up between nature and culture that presumably was not known to Greek or Roman culture.[49]

For this book's concern with Nietzsche's posthumanism, it is important to note that Nietzsche is not relying on a general dichotomy between nature and culture or nature and technology here. Rather, he identifies different cultures and particular modern technologies as detrimental to the nature–culture continuum. In the quote above, Nietzsche seemingly links the separation of knowledge to increased storage capacities, to the availability of an enormous number of concepts presumably through the availability and accessibility of ever larger archives of knowledge. He subsequently targets academia as exacerbating the problem by producing "not at all the free cultivated man but the scholar ... the precocious and up-to-the-minute babbler" (HL, 117). At issue is not a lack of knowledge, but its overproduction as "science," which "seeks to abolish all limitations of horizon and launch mankind upon an infinite and unbounded sea of light whose light is knowledge of all becoming" (HL, 120).

Nietzsche describes the effects of this information overload with reference to Little Red Riding Hood's wolf:

> In the end, modern man drags around with him a huge quantity of indigestible stones of knowledge, which then, as in the fairy tale, can sometimes be heard rumbling about inside him. And in this rumbling there is betrayed the most characteristic quality of modern man: the remarkable antithesis between an interior which fails to correspond to any exterior and an exterior which fails to correspond to any interior—an antithesis unknown to the peoples of earlier times. (HL, 78)

The quantity of unchecked information and the fact that it finds no immediate application has a twofold effect on modern humans (and, according to Nietzsche, especially on Germans): it creates a disjunction between inside and outside, between content and form, whereby form is rejected as convention, disguise, and dissimulation. When Nietzsche subsequently mocks the Germans as the people famed for inwardness, he in fact mocks leading codes the Enlightenment adopted from the humanist tradition: their valuation of inwardness (spirituality, reason) over the social exterior and the body, and of rationality and sensibility over rhetoric and conversational aptitude.[50]

How much Nietzsche's critique of modern culture is informed by technological considerations is still more apparent in his reflections on the corrosive effects of the mass media of his time, the newspaper. They contribute to "modern man" filling and overfilling themselves with "alien ages, customs, arts, philosophies, religions and knowledge" that turn them into "walking encyclopaedias":

> The whole of modern culture is essentially inward: on the outside the bookbinder has printed some such thing as "Handbook of inner culture for outward barbarians." This antithesis of inner and outer, indeed, makes the exterior even more barbaric than it would be if a rude nation were only to develop out of itself in accordance with its own uncouth needs. For what means are available to nature for overcoming that which presses upon it in too great abundance? One alone: to embrace it as lightly as possible so as quickly to expel it again and have done with it. From this comes a habit of no longer taking real things seriously, from this arises the "weak personality" by virtue of which the real and existent makes only a slight impression; one becomes ever

more negligent of one's outer appearance and, provided the memory is continually stimulated by a stream of new things worth knowing which can be stored tidily away in its coffers, one finally widens the dubious gulf between content and form to the point of complete insensibility to barbarism. (HL, 79)

That Nietzsche packages his observations in the neurophysiological discourses of his day highlights how he attributes hominization effects to technology. The way modern communication media technologies process and distribute knowledge affects directly human cognition, and with it gives rise to a particular personality, to new sensibilities (barbarism), and to the self-alienation he describes (the gulf between inner and outer). The media's instant representations of current events "transformed man almost into mere *abstractis* and shadows" (HL, 84) and into spectators:

> Modern man who allows his artists in history to go on preparing a world exhibition for him; he has become a strolling spectator and has arrived at a condition in which even great wars and revolutions are able to influence him for hardly more than a moment. The war is not even over before it is transformed into a hundred thousand printed pages and set before the tired palates of the history-hungry as the latest delicacy. It seems that the instrument is almost incapable of producing a strong and full note, no matter how vigorously it is played: its tones at once die away and in a moment have faded to a tender historical echo. (HL, 83)

Nietzsche lays out how modern media technologies temporalize meaning such that they rob it of its ability to cultivate humans in ways that would connect them to their surroundings. From the perspective of twentieth-century information theory, Nietzsche seems to recognize that science and the mass media process information as news, as something that is only of singular, and hence temporary, significance. News is a "difference which makes a difference"[51] only once, to follow Gregory Bateson's definition of information. As soon as it is encountered a second time, it loses its informational value, it is no longer news, but an old hat. Modern information-processing systems compensate for the chronic loss of information by sustaining a constant flow of information, creating and replacing information with ever new differences that again will leave behind little more than another momentary difference within the system of communication.

Today, we are all too familiar with this mode of information processing and its inflationary effects. As part of the constant barrage of more or less catastrophic "breaking news," the particular content seems to have very little effect on anything, while the sustained flow of information cultivates people in ways that turns them into what Nietzsche called "spectators," abstractions and shadows. That Nietzsche's diagnosis was published 150 years ago is an indication that we are dealing with a legacy that is older than cable news, the internet, or social media, albeit these new technologies further accelerate the malaise that Nietzsche diagnosed: precisely because there is an oversaturation of information, society is only rarely affected in a significant way by any particular event. One wonders whether such information processing is not a kind of immune system in Western society. Just as global media coverage is able to bombard everyone with a continuous stream of more or less catastrophic events, it devalues these events, ensuring that they rarely have a measurable impact on anyone other than those who are immediately affected.

There are apparent parallels and differences between Nietzsche's and Marx's analysis here. Both challenge Enlightenment dogma about education's ability to produce cultivated individuals as the basis for a civil society. Both identify technology as central to the hominization process, as structuring human relations, and with them human sensibilities and thinking. Both lay bare alienation and "barbarism" at the base of modern society. While the early Nietzsche does not engage economics per se or miserable factory working conditions, he does link the "educational operations" to the needs of the labor market:

> That means, however, that men have to be adjusted to the purposes of the age so as to be ready for employment as soon as possible: they must labour in the factories of the general good before they are mature, indeed so that they shall not become mature—for this would be

a luxury which would deprive the "labour market" of a great deal of its workforce. (HL, 97–98)

Nietzsche even recognizes how capitalism ruins education and science. Still in the same section he laments:

> I regret the need to make use of the jargon of the slave-owner and employer of labour to describe things which in themselves ought to be thought of as free of utility and raised above the necessities of life; but the words "factory," "labour market," "supply," "making profitable," and whatever auxiliary verbs egoism now employs, come unbidden to the lips when one wishes to describe the most recent generation of men of learning. (HL, 99)

That Nietzsche's concern is primarily with the middle class and not with the exploitation of the working class does not mean that he does not respond to the same comprehensive social and conceptual changes that Marx attributes to the advent of capitalism. Like Marx, Nietzsche recognizes how these social changes have cultivated humans in ways that have subjected them to uniformity and made society uniformly less civilized, rather than more civilized. Nietzsche supplements Marx's recognition of the effects of machinery by adding pedagogical techniques and communication media to the list of technologies that produce a particular set of modern human beings who experience themselves as alienated from themselves. As we noted earlier for Marx, for Nietzsche too, alienation is not the result of technology or culture compromising a human nature that would exist independent of its cultural, social, and technological embeddedness. It is rather the product of a particular culture and the particular use of specific technologies that create a particular sense of alienation where, as Nietzsche puts it, the "inner" of one's desires, values, and knowledge does not align with the "outer" of one's doings, sense of purpose, of direction, of belonging, and so on. Needless to say, such alienation happens not just under the dire conditions of nineteenth-century factory work or in similarly demeaning work conditions in modern-day sweat shops, computer manufacturing facilities, or mass distribution centers, but also in physically less hazardous settings like retail, fast food, office cubicles, or the vast majority of low-skill service jobs.

Nietzsche's contentions about modern Western "educational operations" and newspapers anticipate twentieth-century media theories (Havelock, McLuhan, Kittler) that connect changes in communication-media technologies to comprehensive changes in communication patterns, semantics, and social relations. McLuhan's assertions about the "galactic" effects of the printing press are as well-known as his laments about the destructive effects of electronic media. Whether we want to agree with his sentimental view or not, what McLuhan's narrative of loss shows is a recognition that the humanist values he sees under threat are the product of the printing press and the emergence of a book culture that ostensibly is coming to an end with the dominance of electronic and digital media in the twentieth and twenty-first centuries. In chapter 1, I quoted Stefan Herbrechter's suggestion that posthumanism is best understood as a reaction to the long tradition of Renaissance book culture coming to an end as new media technologies are replacing traditional reading practices, studying habits, writing patterns, and associated semantics. Not to ignore how contemporary advances in communication-media technologies produce new changes or how there are constant frictions between different media paradigms and the semantics they foster, Nietzsche's analysis of the destructive effects of the mass media of his day seems to move the date of the media-technology-induced end of humanism up by almost a century. According to Nietzsche, humanist values were already collapsing in the second half of the nineteenth century under the weight of the modern educational operations and the effects of the printing press.

On *Übermenschen* and Cyborgs

While it might seem that the importance this chapter attributed to technology for hominization and cultivation processes plays into transhumanist appropriations of Nietzsche's thought, there are important differences. For one, Nietzsche does not theorize technology as an enhancement or prosthesis to human nature, but as intricately linked to the evolution of humans, their sociality, and their sensibilities. Furthermore, Nietzsche is interested in how technologies create, challenge, and change modes of thinking, from Platonic dualisms to the Enlightenment belief in the efficacy of education. The next chapter will examine more closely how he targets in particular insinuations of teleology and "progress" (in the liberal sense) that are essential to transhumanists and futurists in their musing about the advent of a technologically enhanced posthuman. As a consequence, he is reluctant to give a clear contour of even his most obvious model for a future posthuman human, his figure of the overhuman. Nietzsche the "prophetic vitalist" keeps "craving the coming of the overman,"[52] as Herbrechter caricatured transhumanism's Nietzsche, because he recognizes the difficulty of escaping the long history of hominization and the forces it unleashed.

Graham has written the most convincing critique of transhumanism's illegitimate appropriation of Nietzsche's *Übermensch,* challenging the idea that transhumanism represents "a latter-day Nietzschean sensibility."[53] The problem is that transhumanism (like other strands of dystopic and liberal posthumanism, we might add) sees itself "as the twenty-first century heirs to Enlightenment traditions of secular liberal humanism, in which humanity, having displaced the gods, achieves heights of wisdom and self-aggrandizement."[54] The idea that technology could help "perfect" humans—which, not coincidentally, in many sci-fi novels, movies, and games, takes the form of a digitization of the mind—reflects "a deep-rooted philosophical tradition in Western thought, a Platonic world-view in which the physical sensory world is but a reflection of a purer, ideal realm of perfect form."[55] In her critical assessment of transhumanism, Graham adopts Nietzsche's (and Heidegger's) view that connects humanism to the long tradition of Platonic idealism and neo-Platonic (Judeo-Christian) metaphysical thought. Transhumanism, in other words, extends the very tradition of metaphysical and humanist thought that Nietzsche's philosophy challenges at its core.

There are additional problems with the appropriation, and also with the rejection of Nietzsche by dystopic and liberal strands of posthumanism. As this chapter hoped to show, one is the use of a rather narrow notion of technology that is applied in transhumanist and futuristic philosophies without historical reflection. Ironically, because of their narrow definitions of technology as scientific gadgets, instruments, programs, and the like, the most technophile strands of posthumanism fail to recognize the constitutive role technologies have and have had all along for the evolution of humans: physiologically, psychologically, culturally, socially, and politically. Contrary to opposite claims by Kurzweil, More, Stefan Sorgner, and others,[56] Nietzsche is heavily invested not only in the question of technology but also in the reciprocity between human and technological evolution.

I suggested at the outset of this chapter that the cyborg was the emblematic figure for certain strands of posthumanism and transhumanism, but I also pointed out that, at the very moment the cyborg comes to embody the possibility of a merger between technology and the organic, it asserts their heterogeneity: that technology is essentially external to humans. I want to conclude this chapter with a different reading of the cyborg figure, one that we can now assess as more Nietzschean than its transhumanist interpretations would have it be. Haraway's famous "Cyborg Manifesto" sees the cyborg neither as a prosthesis nor as an extension of the human physiology and its functions, nor is she extending Platonic dualisms or metaphysical conceptions of teleology and progress. Instead, she explores how the cyborg figure challenges conceptually existing paradigms of thought and the social hierarchies they support. That is, she explores the *conceptual* impact of the figure, and by extension the conceptual impact of technological innovation, the mental conceptions that flow from technology. For Haraway, then, the importance of the cyborg lies less in proposing a model for the technological perfectibility of humans, and more in how the figure challenges, along with the border between the organic and machinery, what is recognized as nature in the first place. The coupling of technology and humans (as biological entities) thus allows us, Haraway argues, to reconceive and redraw "the territories of production, reproduction, and imagination" and question "the certainty of what counts as nature,"[57] including with regard to questions of gender.

Unfortunately, in the vast majority of popular novels, movies, and video games, cyborgs are rarely employed to redraw borders or challenge what is viewed as natural, but instead replicate and reinforce norms and stereotypes. This is most apparent with regard to gender. Cyborgs from *Blade Runner* to *Ex Machina,* like the voice representations of corporate AIs such as Siri, Alexa, and Google Assistant, reinforce gender norms or serve as a foil to highlight what Western culture considers natural versus unnatural or authentic versus inauthentic behavior. The tension between the political hopes articulated in Haraway's "Cyborg Manifesto" and the reality of popular culture does not refute Haraway's approach, but points toward the intersection between the material and the conceptual that Nietzsche identifies as the locus of a continued political struggle. Far from preaching a technological determinism, Nietzsche's constructivist epistemology can help us recognize how the meaning of technoscientific innovations both shapes and simultaneously is shaped by long chains of competing uses and interpretations. Recognizing the dual movement of technological and (agonistic) conceptual development, Haraway's "socialist-feminism," then, is as much indebted to Nietzsche as to Marx.

The possible continuities between Nietzsche and the early writings of Haraway return us to the question of agency and modern subjectivity and notions of the self and their social and political relevance for posthumanism. Against the backdrop of the understanding of technology I develop in this chapter, the following chapter will examine the particular linguistic, grammatical, and interpretive techniques Nietzsche identifies as having produced what he calls "the sovereign individual" and its modern sensibilities. That is, the chapter will focus on how Nietzsche in his later writings understands language itself as a primary technology of hominization and cultivation.

6 Cultivating the Sovereign Individual

> That everyone as an "immortal soul" has equal rank with everyone else, that in the totality of living beings the "salvation" of every single individual may claim eternal significance, that little gnats and three-quarter-madmen may have the conceit that the laws of nature are constantly broken for their sakes—such an intensification of every kind of selfishness into the infinite, into the impertinent, cannot be branded with too much contempt. And yet Christianity owes its triumph to this miserable flattery of personal vanity: . . . The "salvation of the soul"—in plain language: "the world revolves around me."
> —Friedrich Nietzsche, *The Antichrist*

CHAPTER 1 drew on Tamar Sharon's cartography of posthumanism to explore the great divide that separates strands of posthumanism that hold on to normative ideas of human nature—as something either compromised or enhanced by technology—and strands of critical posthumanism (what Sharon differentiates as radical, methodological, and mediating posthumanisms) that challenge such presuppositions in their attempts to arrive at modes of thinking that have the "potential to usher in a postmodern and post-anthropocentric era."[1] As useful as Sharon's distinction is, the contention that critical strands of posthumanism do not resort to notions of human nature or fully escape anthropocentrism calls for closer examination. After all, many critical posthumanists agree that we cannot simply abandon notions of self, human subjectivity, or agency, but have to reimagine such notions in ways that avoid the normative forces of the humanist tradition. Rosi Braidotti, one of the most prominent critical posthumanists, rejects explicitly the kind of essentialization of human nature we find in dystopic or liberal strands of posthumanism, yet she returns repeatedly to an anthropocentric viewpoint that reaffirms a rational and autonomous concept of the subject in her frequent uses of first-person pronouns as much as in her insistence on the need of "at least *some* subject position" with the ability to "autopoietically self-style" in ways that accord with "what we humans truly yearn for."[2] Even if the object of this yearning is subsequently qualified as an ostensibly Dionysian desire "to disappear by merging into a generative flow of becoming, the precondition for which is the loss, disappearance and disruption of the atomized, individual self,"[3] the argument asserts an inherent human nature (as becoming, as relational) and casts the self as an independent entity just as it claims their dependence on the material, biological, technological, and social structures that they embody and within which they find themselves embedded. Put differently, Braidotti still argues from the position of the self rather than from the perspective of the "generative flows" from which Nietzsche, in line with certain strands of radical and methodological posthumanism, derives formations of self, as this chapter hopes to show.

The challenge to develop a postanthropocentric perspective that does not reintroduce new notions of human nature is also pertinent to this book considering Nietzsche's apparent naturalism, a topic that continues to be the subject of intensive debate in Nietzsche scholarship.[4] If naturalism, as Christian Emden puts it, "at its very core holds that human beings are no special case vis-à-vis the rest of nature and, second, that the way we think philosophically about our position in the world should entertain a close relationship to the natural sciences,"[5] then this book's investigation into Nietzsche's epistemology and the entomological aspects of his concept of sociality contributes to the view that Nietzsche's philosophical anthropology is at heart naturalist. Yet, as noted in chapter 2, such claims are subject to an important caveat, which is the fact that Nietzsche recognized the paradox that our understanding of nature, too, is limited by the bounds "nature" (physiology) and language put on human understanding. Based on the evidence provided by the natural sciences, especially neurophysiology, combined with Nietzsche's reflections on the structure of language, there is no immediate access to nature in and for itself. This insight is important not just epistemologically but also with regard to Nietzsche's treatment of language as being intricately involved in cultivating human sensibilities, and with them how humans of a certain time and age experience their selves and what is "natural." Nietzsche indeed treats language as a primary *technology* of hominization and cultivation, a proposition he explicates most comprehensively in his treatise *On the Genealogy of Morals*. Along with the genealogy of morals, the text also describes the emergence of the modern self or "sovereign individual" (GM II: 1) who has come to embody these morals and accompanying sensibilities. This chapter will turn to this famous text with the aim to trace more closely the posthumanist understanding of language as a technology it offers. By understanding language as a technology of cultivation, Nietzsche challenges both the traditional privileging of humans based on their linguistic abilities and the dichotomy between nature and culture. It is with regard to the latter that Nietzsche's elaboration of the genealogy of the modern humanist subject also offers a productive outlook on notions of a posthumanist subject as discussed in today's critical posthumanist literature.

Self-Styling and Embedded Selves

Before turning to Nietzsche, I want to expand a bit more on the difficulty a critical posthumanism faces that emphasizes embodiment and embeddedness while also seeing the need to retain "some subject position" able to self-style in accordance with basic human needs. If the latter is necessary, it is because it is viewed as the basis for the ethical postulates of a posthumanism that hopes to promote pluralism, equality, openness, mutual respect, and a new sense of community. As Braidotti puts it, the loss of "a unitary definition of the human sanctioned by tradition and customs" also marks a new beginning, since "we do remain human and all-too-human in the realization that the awareness of this condition, including the loss of humanist unity, is just the building block for the next phase of becoming subjects together."[6] I will return to the political challenges such appeals for community present in the next chapter. The question at hand is, simply put, how to carve out selves with agency in ways that respect their biological, social, and technological embeddedness.

In her 2020 book *Philosophical Posthumanism,* Francesca Ferrando turns to Michel Foucault's biopolitical perspective, suggesting that the Foucauldian technologies of production, of sign systems, of power, and of self serve as important "(re)sources" for contemporary posthumanists.[7] She cites Braidotti's argument that the "posthuman knowing subject must be understood as a relational embodied and embedded, affective and accountable entity and not only as a transcendental consciousness."[8] Ferrando summarizes additional examples of posthumanists who emphasize embodiment and embeddedness, such as Karen Barad, who sees agency as emerging "through specific intra-actions,"[9] and Donna Haraway, who in her more recent work draws on the biologist Lynn Margulis, advocating that we replace the idea of autopoiesis with that of "sympoiesis" to highlight the role symbiosis plays in evolutionary processes.[10] Most pertinent for this study, however, is the fact that Ferrando turns to Nietzsche to articulate the "core of Philosophical Posthumanism," arguing that Nietzschean perspectivism "is of key relevance to our understanding of a posthumanist epistemology" (149) *and* posthumanism's underlying ethics: "They both embrace the perspectivist critique to absolute universalism and, within the contemporary debate, support the shift from generalized multiculturalism—which has been criticized from many different perspectives, notably by feminism—to situated pluralism and diversity" (151). Ferrando concludes her reading of Nietzsche's perspectivism with the assertion that: "Pluralism, with its emphasis on the respectful coexistence of different perspectives, individuals, groups, and systems, is at the core of Philosophical Posthumanism" (151).

Ferrando brackets her appeal to a Nietzschean perspectivism with the disclaimer that she does not want to "fully embrace his proposal" (149). She also does not offer a detailed analysis of the epistemological contentions that lead Nietzsche to advocate perspectivism or any reflection on Nietzsche's own ethics. Such omissions are problematic, because putting the "emphasis on the respectful co-existence of different perspectives, individuals, groups, and systems" does not compute easily with the emphasis Nietzsche puts on social contentiousness[11] or his insistence on rank-order,[12] let alone with his concept of the will to power and declarations such as: "What is good? Everything that heightens in human beings the feeling of power, the will to power, power itself" (A, 2).[13] This is not to deny that the affirmation of perspectivism and of situatedness are central to Nietzsche or that a Nietzschean ethics can be derived from it. As Paul Patton has shown, Nietzsche's understanding of moral values as "contingent, historical products" does not mean the abrogation of values, but that their normative force derives from their social situatedness, from "the collective social practices, ways of being and forms of life that are sustained or affirmed in acting in accordance with those rules."[14] Considering, however, the fundamentally agonal structure of Nietzsche's thought, any claim about continuities to the ethics of contemporary posthumanism requires a more careful analysis of the ethical thrust of his philosophy. Such analysis might also serve as a safeguard against articulating ethical ideals that, to a great extent, end up replicating highly humanist modes of thinking about perspectives, individuals, groups, and systems constituting seemingly autonomous and sovereign entities whose insights and good will ("respectfulness") we shall rely on for the good of humanity.

To be clear, the problem is not the desire to promote a situated pluralism, diversity, and the standards of social justice, equality, inclusivity, and humaneness we associate with these ideals in today's world. The question for a critical posthumanism is how to do so without repeating the kind of humanist conceits that are pragmatically suspect and may undermine these ideals. While emphasizing the embodied and embedded nature of our existence can help us break with some of the universalist contentions of the past, issues arise when this approach or the use of some of these concepts continue to presuppose selves or subjects as entities that somehow exist prior to and independent of the environments on which they are thought to depend. To put it more pointedly: too often, the recognition of their environmental dependence remains dependent on the recognition of their independence. Notions of self, consciousness (the awareness of the self's becoming), or the body are juxtaposed to rather than derived from the systems and environments in which they are thought to be embedded. As a consequence, this approach, despite claims to the opposite, neglects to acknowledge fully the constitutive force of the biological, social, and technological assemblages from which selves and their particular affective dispositions, intentions, and reasoning emerge in the first place.

In the preceding chapters, this book turned to the life sciences and to a broader understanding of technology to suggest that Nietzsche offers a more radical understanding of organic, social, and technological embeddedness, one that no longer presupposes a self or transcendental consciousness, or even an individual will, but tries to understand how such notions emerge from the interactions between the different environments within which they are embedded. If, as Manuel Dries suggests, "Nietzschean selves are best understood as complex, embodied systems of drives with affective orientations, as well as embodied unconscious and conscious values,"[15] the focus needs to be on the evolutionary processes and interpretive schemas that Nietzsche identifies as producing these selves, rather than on the self. This also applies to presuppositions of human nature and the body. As useful as Foucault's biopolitical lens is in analyzing the techniques and disciplinary regiments that have helped create and continue to shape modern Western notions of self and subjectivity, Foucauldian approaches tend to conceive, as the basis of the self, a body that is thought, as Judith Butler puts it, as "a surface and a set of subterranean 'forces' that are, indeed, repressed and transmuted by a mechanism of cultural construction external to the body."[16] Doing so, this approach runs the risk of naturalizing the body, neglecting to understand it too as a cultural artifact that is defined, treated, and experienced differently by different cultures. Arguing from a perspective that presupposes a body, self, or subject as an independent entity, even if argued as embedded, in a strict sense thus works against what Ferrando identifies as the core of philosophical posthumanism, the recognition and promotion of a situated pluralism. If the technologies of self that Foucault detailed can serve as a resource for posthumanism, as Ferrando rightfully suggests, one should not ignore the genealogy of the modern self, how it came to separate itself within its environments from its environments. Nor, as Butler reminds us, should one neglect to reflect on the conceptual apparatus, the cultural and historical situatedness of the language and logic that developed in

Western societies and that produced what are understood and experienced as body, self, and subject in the first place.

Language as Technology

If language itself is understood as a technology of hominization and cultivation, then Nietzsche's *On the Genealogy of Morals* offers a unique example of how to think the emergence of the modern self and its sensibilities not from the position of the self, but from the point of view of its environments. We saw Friedrich Kittler suggest as much when he argued that writing, "rather than presenting the subject with something to be deciphered, makes the subject what it is."[17] Kittler's statement was tied to Nietzsche's use of a typewriter, but writing might well be understood in the broader poststructuralist sense as referencing the material, technological, and structural conditions of the use of language. By tracing the technicity of language in Nietzsche's *On the Genealogy of Morals*, I do not want to return to the "baroque intricacies of the linguistic turn,"[18] as N. Katherine Hayles quips in her defense of the new materialisms, or replace the traditional privileging of the category of self-consciousness with a privileging of human language. Rather, the point is to recognize how language in Nietzsche plays an integral role not just epistemologically and politically, setting and reinforcing conventions and with them the social hierarchies and value systems that sustain them, but also as a technology of hominization that is intricately involved in the production of human subjects and their culturally specific sensibilities. This is most evident in *On the Genealogy of Morals*, as each of its three essays describes different stages and the corresponding techniques that Nietzsche sees as instrumental in the creation of the "sovereign individual." The first essay depicts the semantic changes and grammatical structures that make possible the invention of a doer behind the deed and with it the replacement of "noble" with moral values. The second essay focuses on the disciplinary techniques and economic calculus that led to the "internalization of man" (GM II: 16) and the emergence of conscience. The third essay observes the various uses and abuses of ascetic ideals in Nietzsche's own time, where ascetic ideals were no longer viewed as specific to a particular (religious) form of hominization, but came to acquire quite different functions and meanings for different people, in different circumstances and situations.

Nietzsche identifies a grammatical operation, the distinction between subject and verb, as the minimal condition for the possibility of something like a "sovereign individual" to evolve. By separating a doer from the deed, language offers an observational tool that makes appear agency and that subsequently makes it possible to attribute choice, intentionality, responsibility, and so on to the doer. (I am using the awkward terms "doer," "deed," and "doings" to avoid derivatives of the verb "act," as the latter immediately invokes an actor or agent with intentions and plans behind the deed.) Nietzsche pinpoints this grammatical operation in section 13 of the first essay of *On the Genealogy of Morals*. The section starts with the memorable image of the lambs who hate the birds of prey who eat them, while the birds of prey think: "'We don't bear any grudge at all towards these good lambs, in fact we love them, nothing is tastier than a tender lamb'" (GM I: 13). Notwithstanding the macabre joke, the point of the image is to make relatable the resentment of the oppressed and highlight the subsequent operation that allows the oppressed—or more precisely, the "priestly cast" in the name of the oppressed[19]—to take revenge and empower themselves, not in the realm of deeds, but of interpretation. It is here that Nietzsche highlights the technicity of language:

> A quantum of force is just such a quantum of drive, will, action, in fact it is nothing but this driving, willing and acting, and only the seduction of language (and the fundamental errors of reason petrified within it), which construes and misconstrues all actions as conditional upon an agency, a "subject," can make it appear otherwise. And just as the common people separates lightning from its flash and takes the latter to be a *deed*, something performed by a subject, which is called lightning, popular morality separates strength from the manifestations of strength, as though there were an indifferent substratum behind the strong person which had the *freedom* to manifest strength or not. But there is no such substratum; there is no "being" behind the deed, its effect and what becomes of it; "the doer" is invented as an afterthought,—the doing is everything. Basically, the common peo-

ple double a deed; when they see lightning, they make a doing-a-deed out of it: they posit the same event, first as cause and then as its effect. The scientists do no better when they say "force moves, force causes" and such like,—all our science, in spite of its coolness and freedom from emotion, still stands exposed to the seduction of language and has not rid itself of the changelings [*Wechselbälge*] foisted upon it, the "subjects" (the atom is, for example, just such a changeling, likewise the Kantian "thing-in-itself"). (GM I: 13)

The subject, defined here broadly as the doer behind the deed, as the agent that causes an action, is the product of this linguistic operation. By splitting deeds into doer and deed, the medium of language makes it appear as if a subject, a free-wielding agent, existed independent of the deed. As it is petrified in language, this separation is a preconscious schema that structures how any event is perceived. Even when there is no referent, as in the German *es blitzt* that Nietzsche cites, or the English "it rains," where *es* or "it" does not refer to any entity (if it was the clouds, it had to be plural), language requires the introduction of a doer behind the deed. The pronoun is needed to make the sentence grammatically and logically correct. To put it bluntly: Nietzsche sees not subjects, individuals, or even consciousness as authorizing human agency; rather, he recognizes how language serves as a technology that grammatically opens up the semantic space for the creation of notions of agency, self, and subjectivity. Language subsequently guides the development of wide-ranging techniques, practices, and institutions that are based on and extend these notions of agency, making them cognitively and socially durable realities.

Nietzsche suggests that the creative force of language extends beyond the social realm even into the hard sciences. In physics, the search for atoms and ever smaller particles (continued today in exorbitantly expensive particle accelerators) is driven by the seductions of language, especially its need for a subject. Whether in physics or in society, such "subjects," Nietzsche argues, were placed there like a changeling (*Wechselbalg*) by the fairy that is language. In German folklore, the changeling is a demonic creature that is foisted on someone in lieu of a stolen child. Whether in the moral sphere or in science, the search for agents behind events falls for a mythological figure, an illusion, that is the product of grammar, not an adequate reflection of a reality that would exist prior to and independent of the technology (the linguistic operations) that makes it possible to think, perceive, and communicate such agency. Nietzsche's point is not that therefore this subject or agent is not "real," but in line with the epistemological positions traced in chapter 2, that agency, as much as "nature," even in its scientific rendering (atoms, subatomic particles), owes its reality to the structures and technologies that make it appear.

Robert B. Pippin, challenging a tendency in Nietzsche scholarship to naturalize psychological motives, goals, intentions and the like, has spelled out Nietzsche's alternative perspective, which derives intentions and the construal of subject authority from the deed itself. "The deed *alone* can 'show' one who one is."[20] To illustrate this point, Pippin offers the writing of a poem as an example for how we might think of intentions forming only in the process of one's doings. The precise "intentions" of the poet only take shape during the writing process, as the poem comes together in an aesthetically interesting way (or not): "To ask for a better poem is to ask for another one, for the formation and execution of another intention."[21] David Owen relates this "anti-regulist"[22] perspective on human agency to Kant's aesthetics of genius, and thus to the distinction between following predetermined rules and discovering new rules during the creative process. Owen finds in this form of creative agency a model for the sovereign individual and "Nietzsche's expressivist account of ethical agency."[23] The recognition of a new form of creative agency can be linked to larger changes in late-eighteenth-century aesthetics. The demand that art be inventive (original and unique) meant that the artist could no longer follow externally prescribed rules or preexisting models, or that they could fully conceive the artwork before the process of its completion. The inventive quality of the artwork and its coherence henceforth had to emerge in the process of its creation. These expectations led to a decentering of the artist and agency, to which the aesthetic discourses of the time responded with paradoxical notions of artistic genius. When Kant defined genius as an "innate mental predisposition [*ingenium*] *through which* nature gives the rule to art,"[24] he attempted to recuperate the loss of authorial control by naturalizing the genius's talent, by having the genius "unknowingly know" or feel what, in a strict sense, they cannot know in advance for the artwork to be inventive. For the most part, it took until the twentieth century for aesthetic theories to be developed that rejected metaphysical notions of agency and instead focused on how artists attune themselves to the emerging artwork or the performance, reacting to the choices that present themselves during and based on the production process until the work achieves a level of saturation or cohesion that invites an end.[25] In this regard, the notion of genius shows the stubbornness with which the insinuation and presupposition of agency holds even in processes that must by definition challenge the authority and independence of the doer.

Extending Pippin's and Owen's reading of Nietzsche's modeling of agency beyond the arts makes it possible to appreciate more how much "doers" emerge as part of their technological situatedness. Different doers or selves will derive from different, often intersecting doings. Situatedness conceptualized along these lines does not necessarily mean that there is no stability to agency, perceived, behavioral, or in style. As the persistence of the figure of artistic genius shows (used to this day at least for the purpose of marketing), there is no lack of interpretive means, and by extension, of practices and institutions (banks, the courts, schools, voting booths, and so on), that constrain the doers, but also lend them and their intentions a certain degree of durability, consistency, reflective distance, and reach. But this only underlines how selves, subjects, and agents are the product of the interpretive operations and practices with which they engage, and in which they are embedded, in their doings.

What makes Nietzsche's argument productive for posthumanist considerations of agency is that it articulates a notion of embeddedness that challenges radically any thinking that locates a self, subjectivity, or agency in opposition to the social and technological (including linguistic) forces from which it emerges. Without the technicity of language, no sovereign, coherent, or autonomous human could be recognized, and no selves could think of themselves as shaping their surroundings from the outside. Nietzsche's assertion of the primacy of language as a technology with which selves, subjects, and human thoughts and relations can be construed, structured, maintained, and/or challenged is in line with that of Ernst Kapp, whose *Elements of a Philosophy of Technology* defines language as "the tool through which it understands itself as its own tool—a spiritualized tool, both apex and agent of the human being's absolute self-production."[26] In recognizing the technicity of language—as Kapp puts it, "language thinks in us and speaks out from us"[27]—Nietzsche anticipates critical posthumanist contentions that, following Cary Wolfe's assessment, are flushed out fully only by those whom Wolfe identifies as most "exemplary posthumanist theoriest," Jacques Derrida and Niklas Luhmann. If Derrida and Luhmann are most exemplary, it is "because both refuse to locate meaning in the realm of either the human or, for that matter, the biological."[28] The displacement of meaning away from the human mind derives from their insistence on the heterogeneity of speech and self, communication and consciousness, and the priority they attribute to the medium, the non-individual, socially constituted technology of communication over the self-presence of speech and the insinuation of medial transparency. The necessary reliance of thought, perception, and even feeling on the technicity of signs, language, and communication led both Derrida and Luhmann to articulate theoretical positions that recognize, Wolfe suggests, "that the human is, at its core and in its very constitution, radically ahuman and constitutively prosthetic."[29] The latter term might be a bit deceiving, as "prosthetic" invokes a notion of technology that would extend or supplement human nature. What is at stake here, however, is the *constitutive* role of language, how this technology must be thought to be inseparable from what it helps construct. As *On the Genealogy of Morals* shows, Nietzsche conceives of language as a primary technology of hominization that is, as Wolfe puts it pointedly, "a radically ahuman precondition for our subjectivity."[30]

Nietzsche's deconstruction of the autonomy of the subject must seem to offer an attractive alternative to Kant's stipulation of a transcendental consciousness, a regular target of critique among posthumanists. Yet, as the brief excursion to Kantian aesthetics above suggested, there might be more continuity between Kant and Nietzsche than first meets the eye. John Richardson has made this point specifically with regard to what he summarizes as Nietzsche's main (albeit not always compatible) positions that the subject does not exist and that agency is an epiphenomenon, is harmful, or is a mere tool of the drives.[31] While these concepts challenge the idea of a "transcendental consciousness," Richardson is able to show how Nietzsche's skepticism regarding the subject as agent is not contrary to Kant, but rather an extension and radicalization of Kant's own thinking. Both Kant and Nietzsche take "from Hume the lesson that we can never have evidence of a subject, and that it's an error to treat the I as a substance. But [Kant] also agrees with Nietzsche's (sometime) view that taking-oneself-to-be a subject is necessary since the 'unity of apperception' is a condition for any experience. Kant agrees, moreover, that we can never have evidence of an agent acting from a free choice among reasons. But he thinks that this posit, too, is necessary as a condition not for experience but for acting."[32] That Kant derives the subject transcendentally, as a possibility condition, suggests precisely *not* that this subject exists independent of nature or as its master, but that it is a feeble construct designed to create a sense of coherence of experience and action where there is none absent of this presupposition. At the very least, Kant's transcendental stipulation does not prevent, but rather invites, a further examination of the physiological, social, and technological processes that necessitated and produced the stipulation of the subject. Nietzsche's analysis makes clear that we need not per se reject or declare as boogeyman the transcendental subject. The point is to recognize as historically contingent the discourses, institutions, and practices from which modern notions of subjectivity and associated sensibilities emerge, as well as the social and political efficacy of these constructs.

The Genealogy of Conscience

It is one thing to argue that language is the technology that makes possible the emergence and conception of a subject-agent, but another to trace how this subject-agent evolved over time and acquired the kind of emotional depth and conscientiousness it seems to be able to harbor (and so often has failed to harbor) since the eighteenth century. In the second essay of *On the Genealogy of Morals*, "On Guilt, Bad Conscience, and Related Matters," Nietzsche expands on the emergence of what he calls the "sovereign individual," this modern, sensible, and sensitive human that has the right to make promises. Nietzsche puts the right to promise at the center of his definition of the sovereign individual. As an exemplary speech act, the promise seems to demonstrate a unique agency of individuals, identifying human actors as center and source of societal formations (think marriage vows). Indeed, in social-contract theories from Hobbes to Locke and Rousseau, the promise was thought to motivate the transition from the state of nature to a civil society that represents and protects the rights of its citizens. Twentieth-century speech act theory as devised by J. L. Austin or John Searle still thinks the performative aspects of language in this way, that is from the position of the actor. Nietzsche, however, puts the "*prerogative* to promise" (GM II: 2) not at the beginning of the modern state, but at the end of a long cultivation process, highlighting how many things had to occur for "this necessarily forgetful animal" to acquire the ability and right to make promises:

> This necessarily forgetful animal, in whom forgetting is a strength, representing a form of *robust* health, has bred for himself a counter-device, memory, with the help of which forgetfulness can be suspended in certain cases,—namely in those cases where a promise is to be made: consequently, it is by no means merely a passive inability to be rid of an impression once it has made its impact, nor is it just indigestion caused by giving your word on some occasion and finding you cannot cope, instead it is an active *desire* not to let go, a desire to keep on desiring what has been, on some occasion, desired, really it is the *will's memory*: so that a world of strange new things, circumstances and even acts of will may be placed quite safely in between the original "I will," "I shall do" and the actual discharge of the will, its *act,* without breaking this long chain of the will. But what a lot of preconditions there are for this! In order to have that degree of control over the future, humans must first have learnt to distinguish between what happens by accident and what by design, to think causally, to view the future as the present and anticipate it, to grasp with certainty what is end and what is means, in all, to be able to calculate, compute—and before they can do this, humans themselves will really have to become

reliable, regular, necessary, even in their own self-image, so that they, as someone making a promise is, are answerable for their own *future*! (GM II: 1)

Rather than presupposing the subject as a sovereign, coherent, and autonomous self, Nietzsche elaborates the disciplinary regiments that make it possible for this forgetful animal to act as a coherent self in the first place. Despite the biological metaphor Nietzsche employs, we should not separate the process from the product here. The technological conditions must remain in place for the possibility of sovereignty. In this regard, as Herman Siemens put it, "Nietzschean sovereignty is non-sovereign in the sense that it depends on cultivating certain relations with others; it is deeply embedded and thoroughly relational in character."[33] Recognizing the embeddedness and technological contingency of the sovereign individual—that it is and remains intricately linked to the humanist modes of hominization that produce it, that "the sovereign individual is, so to speak, the star pupil of the morality of custom"[34]—also suggests that Nietzsche did not think of the sovereign individual as an ideal for emulation or as a possible model of the *Übermensch* or a posthumanist human. When Nietzsche calls the sovereign individual the "ripest fruit on the tree of knowledge" (GM II: 1), the metaphor implies that the sovereign individual stands at the very beginning of a new process, of a new tree that presumably will be cultivated differently.[35]

The subsequent sections of the second essay in *On the Genealogy of Morals* focus on the particular techniques and interpretive schemas that gave humans their "conscience" and modern sensibilities. Nietzsche famously argues that the modern self is the product of a violent past, of penal codes and practices that applied gruesome amounts of pain to the human body:

> We Germans certainly do not regard ourselves as a particularly cruel and hard-hearted people, still less as particularly irresponsible and happy-go-lucky; but you only have to look at our old penal code in order to see how difficult it was on this earth to breed a "nation of thinkers" (by which I mean: *the* nation in Europe that still contains the maximum of reliability, solemnity, tastelessness and sobriety,

qualities which give it the right to breed all sorts of European mandarin). These Germans made a memory for themselves with dreadful methods, in order to master their basic plebeian instincts and the brutal crudeness of the same: think of old German punishments such as stoning (—even the legend drops the millstone on the guilty person's head), breaking on the wheel (a unique invention and speciality of German genius in the field of punishment!), impaling, ripping apart and trampling to death by horses ("quartering"), boiling of the criminal in oil or wine.... —Ah, reason, solemnity, mastering of emotions, this really dismal thing called reflection, all these privileges and splendours man has: what a price had to be paid for them! how much blood and horror lies at the basis of all 'good things'! .... (GM II: 3)

Nietzsche dubs this process, "the ever-growing intellectualization and 'deification' of cruelty, which runs through the whole history of higher culture (and, indeed, constitutes it in an important way)" (GM II: 6). He concludes memorably: "All instincts which are not discharged outwardly *turn inwards*—this is what I call the *internalization* [*Verinnerlichung*] of man: with it there now evolves in man what will later be called his 'soul'" (GM II: 16).

Today, we read Nietzsche's cultivation theory through Foucault, who, in his 1975 *Discipline and Punish,* provided the historical data for the social changes Nietzsche describes while embedding them in a historical narrative that looks at symbolic, discursive, and institutional shifts that accompany the advent of modernity around 1800. Foucault showed how Nietzsche's genealogy registers the sweeping changes in the penal codes of most Western countries, when in the span of a few decades in the eighteenth century, torture and other forms of public punishment that targeted primarily the body were replaced by less overtly cruel practices of punishment that targeted what Foucault calls the modern soul. Foucault does not resort to a causal explanatory model that would posit that the cruel "mnemotechniques" of the past created the modern soul. He speaks of a redistribution and refinement of power through disciplinary regulations that would soon permeate many areas of modern society (not just prisons, but also factories, schools, barracks, hospitals, and so on). In this context, Foucault defines discipline explicitly as a "technology":

'Discipline' may be identified neither with an institution nor with an apparatus; it is a type of power, a modality for its exercise, comprising a whole set of instruments, techniques, procedures, levels of application, targets; it is a 'physics' or an 'anatomy' of power, a technology. (215)

To the contemporary reader, the link Nietzsche establishes between physical pain and cultivation must seem suspect. It is, in fact, reminiscent of outdated pedagogical dogmas that rely on corporal punishment to "force" children and students to learn. It also ignores that it is neither biologically nor psychologically possible to hand down such behavioral modifications to future generations as Nietzsche's *On the Genealogy of Morals* at least on the surface seems to suggest. What are handed down, however, and what are more durable than the example set by the excessive suffering of publicly executed criminals, are interpretation schemas and practices and institutions that build on them, and that thus continue to enable and restrict behavior and communication possibilities. What are handed down, in other words, are the cultivation technologies that produce modern selves and their sensibilities.

This applies also to the increases in memory and reflective capacities that Nietzsche suggests are required for the creation of the modern soul. They are the result of the technological innovations the previous chapter discussed: the printing press, increased literacy rates, the build-up of civil services and other institutions, increasing efforts toward mass education, and the "daily reminders" offered by newspapers and the like. Looking at the hominization effects of technology, it is not necessary, then, to assume that pain and cruelty affect generational changes in human sensibilities, behavior, or even biology (suffice it to say, the twentieth and twenty-first centuries still have plenty of pain and cruelty to offer); rather, the historical evidence suggests that new technologies and the social practices and institutions they enabled created a lasting, externalized "memory," and made, to use Bruno Latour's formulation, society and the behavioral patterns they enforce increasingly durable.[36]

Durable, of course, does not imply everlasting or unchanging. In fact, focusing on the effects of technologies on the hominization process allows us to think about the cultivation of humans without having to resort to simplistic and today largely discredited dichotomies such as natural–unnatural, civilized–uncivilized, educated–uneducated, and the like, dichotomies that naturalize human beings in one form or another. A posthumanist perspective that recognizes the constitutive role of technologies will not attempt to separate a human essence from the techniques and technologies with which human beings intra-act, but will instead try to understand better what practices are at play and how they inter- and intra-act in defining changing notions and experiences of self and subjectivity. It will also try to understand, as Nietzsche does in the second essay of *On the Genealogy of Morals,* potential problems and contradictions inherent to these processes and the subjectivities that emerged from them.

Manas: The Calculating Animal as Such

In the second essay of *On the Genealogy of Morals,* Nietzsche puts at the center of the cultivation process that produced the modern, humanist subject the adoption and judicial application of a basic economic calculus. Expanding on the etymological proximity of the German words for debt (*Schulden*) and guilt (*Schuld*), he links the "*high* degree of humanization . . . so that animal 'man' could begin to differentiate between those much more primitive nuances 'intentional,' 'negligent,' 'accidental,' 'of sound mind,' and their opposites" (GM II: 4) to the "idea of an equivalence between injury and pain," an idea that gained its power "in the contractual relationship between *creditor* and *debtor,* which is as old as the very conception of a 'legal subject'" (GM II: 4). Put in a nutshell, Nietzsche argues that humans punish not to "correct" behavior or to prevent crime, as we like to claim to this day (despite plenty of evidence to the contrary) so as to justify this brutish and ineffective practice, but because humans derive a sense of pleasure, and hence compensation, from seeing suffering. The "payback," as the English idiom goes, lies in the feeling of superiority we get to experience when we punish or see someone being punished. Within the genealogy of morals, it is this psycho-economic calculus that drives the history of punishment, including many forms of self-punishment.

Nietzsche indeed puts this mode of thinking at the very beginning of the evolution of humanity. He speculates that the practice of humans separating themselves and putting themselves above the animal might have its origins in this "most primitive kind of cunning," which "was also, presumably, the first appearance of human pride, man's sense of superiority over other animals" (GM II: 8). He cites the Sanskrit word *manas*, which means both human and mind, as evidence: "Perhaps the word 'man' (*manas*) expresses something of *this* first sensation of self-confidence: man designated himself as the being who measures values, who values and measures, as the 'calculating animal as such'" (GM II: 8). Nietzsche subsequently locates this basic economic calculus also at the "beginnings of any social form of organization or association," arguing that "the germinating sensation of barter, contract, debt, right, duty, compensation was simply *transferred* from the most rudimentary form of the legal rights of persons to the most crude and elementary social units" (GM II: 8). He cites as evidence for the idea that "the community has the same basic relationship to its members as the creditor to the debtor," the practice of older societies who punished criminals for injuring the community by returning them "to the savage and outlawed state" (GM II: 9). Nietzsche continues that, as "the power and self-confidence of a community grows, its penal law becomes more lenient [and the] 'creditor' . . . becomes more humane," to the point where:

> It is not impossible to imagine a society *so conscious of its power* that it could allow itself the noblest luxury available to it,—that of letting its malefactors go *unpunished*. 'What do I care about my parasites,' it could say, 'let them live and flourish: I am strong enough for all that!' . . . Justice, which began by saying 'Everything can be paid off, everything must be paid off,' ends by turning a blind eye and letting off those unable to pay,—it ends, like every good thing on earth, by *sublimating itself.* The self-sublimation of justice: we know what a nice name it gives itself—*mercy*; it remains, of course, the prerogative of the most powerful man, better still, his way of being beyond the law. (GM II: 10)

While giving humans a sense of separation from the animal world, the economic calculus that subtends the civilization process in reality attests to the continued bestiality of humans, the pleasure the "calculating animal" must derive from the infliction of pain to want to accept it as a form of compensation. For Nietzsche, then, the humanist subject, with its purported humaneness, remains at base the product of a long history of violence directed against others and itself; or as he famously puts it toward the end of the second essay: "We moderns have inherited millennia of conscience-vivisection and animal-torture inflicted on ourselves: we have had most practice in it, are perhaps artists in the field, in any case it is our *raffinement* and the indulgence of our taste" (GM II: 24).

In his recent article "Histories of Violence," Emden has shown how and why the importance Nietzsche "clearly attributes to violence and cruelty as anthropological constants and concrete manifestations of power inevitably renders problematic the assumed primacy of the morally good" (228). Nietzsche challenges both Kantian and utilitarian accounts of punishment: "Even the most rationally retributivist conception of punishment cannot but highlight that punishment always entails a violent manifestation of power" (224). The internalization of violence and cruelty that Nietzsche describes as part of the Western cultivation process, "to be sure, allows us to ignore the reality of violence, but Nietzsche understands that it would be disastrous to therefore conclude that the emergence of moral conscience is completely detached from concrete forms of violence" (227). Emden spells out the upshot of Nietzsche's analysis, that it "demands of us to accept the realities of violence without justifying the use of violence" (226). In this vein, it is important to recognize that the psycho-economic calculus Nietzsche locates at the heart of the Western cultivation process is not abolished even when it expresses itself in a nonviolent act such as the granting of mercy. As the Nietzsche quote above makes clear, mercy is uttered from a position of privilege that it affirms when the privileged refrain from inflicting pain. Extending "mercy" is not expressive of a posthumanist ethos that would have overcome the calculus at the heart of the humanist tradition; rather, it sublimates the economic calculus into an essentially patriarchal gesture where the merciful derive pleasure from putting themselves above others and, in this case, also above the law. In this regard, mercy remains, in Nietzsche's idiom, a reactive feeling. Mercy thus should not be equated with justice. The pursuit of justice emerged, as Lawrence Hatab summarizes the argument of sections 10 and 11 in *On the Genealogy of Morals*, "as a battle waged by active forces 'against reactive feelings,' by types who 'expended part of their strength in trying to put a stop to the spread of reactive pathos, to keep it in check and within bounds, and to force a compromise.'"[37]

The ethical and political implications of Nietzsche's analysis—which, as Hatab argues, is in fact why Nietzsche's philosophy is important for democracy—are the subject of the concluding chapter of this book. For now, it is enough to note that Nietzsche's critique of morality targets a discursive practice that allows for the continued justification of violence, whether perpetrated on others or on oneself. Even ideals with "nice names" like "mercy" extend rather than overcome modes of human, all too human thinking and valuing that promote rather than overcome violence. This also applies to institutions of justice that higher powers employ to combat resentment and to control the violence at the heart of the rudimentary economic calculus from which *manas* emerged. By virtue of their target, they remain beholden to the compensatory logic they combat. Emden puts it succinctly: "Morality and law seek to avoid violence by doing violence" (227). A posthumanist society worthy of this name would have to emerge from and subtend institutions, social relations, and modes of thinking that are no longer driven by this archaic and ultimately violent compensatory logic.

It is difficult to fathom the transition to a society that would no longer apply a compensatory logic but would indeed "distrust all in whom the impulse to punish is powerful" (Z II: "On the Tarantulas").[38] It is especially difficult to fathom considering how much human suffering is the product of extensive inequalities of means and voice (educational, material, medical, monetary, political, and so on). With Nietzsche, it is possible at least to explain this difficulty, as such a society would no longer be "human" in the sense this term remains used today, as it would no longer revolve around *manas*, the calculating animal. Nietzsche's analysis can also serve as a warning about humanist pleas for a more rational or more merciful administration of the law and of political power that upholds a status quo that reaffirms a basic inequality as part of its human, all too human calculus.

By focusing on linguistic operations and the evolution of particular interpretive schemas, Nietzsche puts forward a more radical notion of embeddedness of humans and human agency, of self and subjectivity, than can be found in the writings of many posthumanists today. Nietzsche recognizes, to use a more contemporary vocabulary, how the brain is "encultured" as much by material as by symbolic technologies.[39] Acknowledging the constitutive role of the latter asserts both the durability of identity constructions and their openness to change through the kind of critical and creative discursive practices that are central to Nietzsche's philosophy. In this regard, Nietzsche advances a critical posthumanism that is more than an extension or broadening of the sort of pluralism that fails to question the schema of "the human" and can easily be "appropriated for the ideological work of the neoliberal order, in which capitalist globalization gets repackaged as pluralism and attention to difference."[40] Selfhood, as Hatab puts it succinctly, "for Nietzsche, is always emergent within a dynamic of life forces that will disallow any impulse toward 'identity.'"[41] These forces, I would like to add, should not be viewed independently of the linguistic structures, modes of observation, and dynamics of the social and technological environments from which selfhoods emerge and can be expected to change continually.

Toward a Posthumanist Naturalization of Technology

Based on the preceding reflections on the technicity of language, I want to return to Vanessa Lemm's affirmative biopolitical reading of Nietzsche in her book *Homo Natura* and to her critique of the "assemblage posthumanism" she finds in the works of Braidotti and Wolfe as laid out in the book's final chapter, "Posthumanism and Community of Life." Lemm's main point of contention with Braidotti and Wolfe—and where she locates a major incongruency between the two contemporary posthumanists and Nietzsche—is the role they attribute to technology. By understanding technology as a supplement and prosthesis, Lemm argues, both see technology as complementing and extending human nature. In doing so, they adopt a philosophical anthropology that, following Helmuth Plessner's famous observation, defines humans as deficient beings (as

Mängelwesen). According to Lemm, this point of view is incompatible with Nietzsche's naturalism. "For Nietzsche, the problem of civilization and technology is that they immunize the human being against its own animal and plant nature, separating the human being from the community of life and hence also from those drives and instincts that would otherwise allow it to meaningfully engage with other forms of non-human life" (173).

In her discussions of Braidotti's insistence of the "ubiquity of technological meditation," Lemm also discovers an apparent paradox:

> The foundational claim of this assemblage posthumanism, according to which human nature is rejected in favour of "the posthuman subject as a composite assemblage of human, non-organic, machinic and other elements," is the transcendental fact of "the structural presence of practices and apparati of mediation that inscribe technology as 'second nature.'" In other words, and paradoxically, the human can enter into community with "first nature" only by separating itself from it through technology as "second nature."[42]

Lemm's critique points toward a larger issue in Braidotti's writings, which is her attempts to renaturalize the posthuman subject. Although Braidotti seems to favor an antifoundational and constructivist stance rhetorically and in what amounts to her quite Nietzschean aestheticization of politics as being about creativity and openness, she explicitly distances her strand of posthumanism from poststructuralist presuppositions: "The posthuman subject is not postmodern, because it does not rely on any anti-foundationalist premises. Nor is it deconstructivist, because it does not function within the linguistic turn or mediation."[43] By rejecting the poststructuralist legacy, and by replacing the biopolitical with a "conceptual frame of nomadic becoming" in a subsequent essay,[44] Braidotti returns to an essentialized notion of nature (*zoë* understood in the Spinozist tradition as a productive force extended to all "vital matter") that reveals itself, as Lemm notes, in the technology-dependent formation of the posthuman subject. Technology thus remains essential for the appearance of nature, and yet secondary to it.

I have shown in this chapter and the previous one how an expanded notion of technology more appropriate for the nineteenth century presents an alternative to the dichotomy between nature and technology that continues to dominate much of transhumanist and posthumanist thinking on the matter. It can also help resolve the logjam between Lemm's dismissal of technology in her assessment of Nietzsche's posthumanism and Braidotti's paradoxical affirmation of technology for the purpose of a renaturalization of posthuman subject formations. The way out requires that any account of nature includes the technology that makes possible such observations about nature or about what appears as natural in the first place. Braidotti does as much when she argues "in favour of a nature–culture continuum which stresses embodied and embrained immanence and includes negotiations and interactions with bio-genetics and neurosciences, but also environmental sciences, gender, ethnicity and disability studies."[45] After all, to affirm that gender, ethnicity, and disability *studies* contribute to today's understanding of the nature–culture continuum of subject formations is to acknowledge that linguistic operations determine human relations and what appears as nature culturally and scientifically.[46]

Acknowledging the technicity of language also does not necessitate a denial of Lemm's focus on the seemingly foundational ways that Nietzsche employs notions of nature and life: as eternal becoming, as irrational drives, as excess, as will to power, as self-overcoming, and so on. We may well *know* with Nietzsche that it is the "unknowability of drives that grants the transformative capacity of the human being."[47] But based on our findings in chapter 2, we should also acknowledge that Nietzsche is well aware of the paradox (of recursion) such unknowability proclamations entail. Lemm's reading of Nietzsche's *homo natura* opens the door to think of an affirmative coevolution of nature and technology along the lines outlined in this chapter when she argues that, because Nietzsche conceived of nature as a "creative and artistic force" capable of transforming humans, his "reconstruction of human nature is thus not a 'return to nature' but an elevation of the human being through the recovery of the 'more' of nature, of nature's generative and creative force."[48] Any such force will have to be channeled and will be channeled differently based on the social and technological environments in which it finds itself. Defining nature as a generative force leaves open the questions of what directs this force, how it takes shape, how creative it can be, and so on. In focusing so much of his writing on the formative and transformative power of social dynamics and technologies (including language), Nietzsche acknowledges their constitutive role in how nature's generative and creative force appears in humans and elsewhere. In other words, if nature (or life, for that matter) is understood with Nietzsche along these lines, it is no longer necessary to presuppose human nature as "a surface and a set of subterranean 'forces' that are, indeed, repressed and transmuted by a mechanism of cultural construction external to the body,"[49] as Butler suggests Foucault does following Nietzsche; nor does it mean treating "technicity" as something that "always already seeks to enhance"[50] the functionality of humanity, as Lemm characterizes transhumanist and assemblage notions of technology. Instead, the concept of nature as force directs the attention toward how a time and culture constructs what it references and ignores, treats and mistreats, values and devalues as (human) nature. As my readings hoped to show regarding the hominization effects Nietzsche attributes to technologies, human nature appears and can be observed and experienced only in relation to the technologies, practices, and interpretive contentions that surround and constantly redirect and redefine it. Such technologies include the effects of critical history in redefining what is experienced as "first nature," the effects of the modern educational operations and the processing of information as news on cognition and human civility, and the effects of grammar and psycho-economic calculations on the genealogy of the sovereign individual.

To be sure, Nietzsche's rhetoric often makes it difficult not to think of the processes of cultivation he describes as violating in some essential way some core aspect of human nature or life. Yet, despite Nietzsche's harsh condemnations of the irrationality, vindictiveness, and violence at the heart of the Western cultivation processes, it is important to situate these observations within his larger genealogical argument. As dismissive as he is of "slave morality" when he introduces it in the first essay of *On the Genealogy of Morals*, Nietzsche also makes clear that these values became dominant and are what subsequently gave humans depth, making them more interesting. Likewise, despite all the vitriol hurled against Christian conscience-vivisection, Nietzsche contends that bad conscience "as true womb of ideal and imaginative events, brought a wealth of novel, disconcerting beauty and affirmation to light, and perhaps for the first time, beauty *itself*" (GM II: 18). Put differently, that our past now appears as "unnatural" and torturous is itself a consequence of the unfolding and sublimation of a particular mode of thinking that by the end of the nineteenth century had become, we might say, self-aware. As the third essay of *On the Genealogy of Morals* also makes clear, the ascetic ideals of the religious tradition do not simply disappear when the religious beliefs that linked them to a person's salvation seem no longer plausible. They merely have acquired new and different functions under different circumstances. What at first glance might look like a repression and transmuting of an authentic force or bodily existence in Nietzsche's analysis reveals itself as part of a genealogy that made possible such an assessment in the first place, and with it the ability to contemplate a different genealogy for the future. This posthumanist future can be imagined to be constituted by alternative, less violent, less normative, more creative, more self-reflective, and more open cultivation processes based on new interpretive schemas and communication possibilities that might acquire durability in new institutions that no longer extend the psycho-economic calculus Nixetzsche detects at the center of the humanist cultivation process.

The next chapter will address further the ethical and political thrust of Nietzsche's posthumanism. In the absence of having recourse to nature or human nature as a given or even as a given good, we will follow Nietzsche's suggestion that the "best we can do is to confront our inherited and hereditary nature with our knowledge, and through a new, stern discipline combat our inborn heritage and implant in ourselves a new habit, a new instinct, a second nature, so that our first nature withers away" (HL, 76).

7

The Ethics and Politics of Nietzschean Posthumanism

CENTURY-OLD QUARANTINE.—Democratic institutions are centers of quarantine against the old plague of tyrannical desires. As such they are quite useful and quite boring.
—Friedrich Nietzsche, *The Wanderer and His Shadow*

What I'm trying to say is: the democratization of Europe is at the same time an involuntary exercise in the breeding of tyrants—understanding that word in every sense, including the most spiritual.
—Friedrich Nietzsche, *Beyond Good and Evil*

NIETZSCHE is not a political thinker in the traditional sense. Yet, discussions of the political implications of his philosophy have only continued to grow since Herman Siemens and Vasti Roodt noted in the introduction to their expansive 2008 anthology *Nietzsche, Power and Politics* that "it is no exaggeration to say that Nietzsche's significance for political thought has become the single most hotly contested area of Anglophone Nietzsche research."[1] Eleven years later, Siemens and James Pearson, introducing their collection of essays on *Conflict and Contest in Nietzsche's Philosophy,* identified two opposing tendencies in the "political interpretation and appropriation of Nietzsche's thought."[2] On the one side, there are those who read Nietzsche as "a ruthless warmonger inciting his readers to a proto-Fascist war of eradication and oppression;"[3] on the other side we find scholars who draw on Nietzsche in support of democratic and liberal values. Both camps are further differentiated. The first camp falls itself into two opposing fractions: agitators on the far right who enlist Nietzsche's philosophy for fascist and neofascist purposes and liberal thinkers who see the appropriation of Nietzsche's philosophy by the far right as a profound danger. The latter feel an urgency to warn about Nietzsche and about those who wish to "domesticate Nietzsche's ideas concerning class and caste, race and sexuality, and his opposition to forms of liberalism, democracy, feminism and socialism."[4] Both arguments have a long history. Attempts to appropriate Nietzsche for fascist and protofascist ideologies can be traced all the way back to Nietzsche's own lifetime and even to his own family. They seem undeterred by "Nietzsche's hatred of 'Jew hatred,'"[5] his rejection of nationalism, and his extensive critique of German culture.[6] More recently, agitators on the far right—Ronald Beiner discusses in particular Aleksandr Dugin (advisor to Vladimir Putin), Julius Evola, and Richard B. Spencer—have invoked Nietzsche as a source of inspiration in a world that, since Donald Trump and the successes of analogous political parties in Europe, has been shown to be "populated by far more neofascists than we may have imagined until quite recently."[7] The renewed critique of Nietzsche by liberals and thinkers on the left also follows a long tradition. Jürgen Habermas argued decades ago that Nietzsche's philosophy helped derail the ambitious goals of the Enlightenment tradition to create a more humane, just, and peaceful society.[8] Publications such as Malcolm Bull's *Anti-Nietzsche* or Beiner's *Dangerous Minds: Nietzsche, Heidegger, and the Return of the Far Right* continue to popularize the idea that Nietzsche's philosophy is proto-fascist and potentially dangerous for modern democracies. Of course, as Paul Patton notes in his "Recent Work on Nietzsche's Social and Political Philosophy," warning about the dangerousness of Nietzsche "is hardly news—Nietzsche drew attention to it in *Ecce Homo* (1888)."[9]

It is against the backdrop of this contentious debate that this chapter will reflect on the political implications of Nietzsche's posthumanism. Rather than try to rescue Nietzsche from his fascist appropriations or his liberal critics, however, I want to examine more closely in what respect Nietzsche's posthumanism offers an outside perspective on these political positions. How do Nietzsche's and today's posthumanisms present new outlooks, a new ethos, and along with it, a new thinking about politics beyond the triptych of a world divided into communism, liberalism, and fascism? How might a posthumanist perspective contribute to understanding why the humanist tradition has failed to live up to many of its most cherished ideals, including the creation of a society that would promote equality, individual freedom, and human dignity more successfully? As indicated in the previous chapter, the question gains in significance for this inquiry into Nietzsche's posthumanism because Nietzsche's rejection of Judeo-Christian morals and his well-known affirmations of power and structures of domination seem to contradict a contemporary posthumanist ethos that puts the "emphasis on the respectful co-existence of different perspectives, individuals, groups, and systems"[10] or that is viewed as expressing "a grounded form of accountability, based on a sense of collectivity and relationality, which results in a renewed claim to community and belonging by singular subjects."[11]

Beyond Liberal and Dangerous

Before turning to the scholarly focused discussions of the ethopolitical significance of Nietzsche, I want to address the recent resurfacing of Nietzsche's epistemology as a matter of public contention. In an age that the media have come to describe as "post-truth," not only do champions of liberalism once again find in Nietzsche dangerous affirmations of "cultural norms that are 'anti-liberal to the point of malice'";[12] they also argue that his epistemology has "left us vulnerable to harsh new ideologies that appear to regard respect for truth as a snare for the strong set by the weak."[13] Such contentions do not differentiate the various conceptions of truth Nietzsche discusses, or they would have noted, as argued in chapter 2, that Nietzsche does not deny the ability to distinguish lies or metaphor from truth, but only a particular philosophical conception of truth (Aquinas's) that, by the nineteenth century, was scientifically no longer tenable. The accusation of "relativism" likewise ignores how Nietzsche's philosophy recognizes the rigidity and binding nature of truths as conventions. It is worthwhile repeating that a constructivist epistemology is not tantamount to an "anything goes" relativism, as they are often being conflated. This would mean to ignore the multiple ways in which statements, actions, and even drives and desires are embedded semantically, grammatically, logically, socially, technologically, institutionally, and so on. Bruno Latour's distinction between a relativist and a relationist perspective underscores the difference. In a relationist analysis, he writes, "all statements are not equal. It is relationist: showing the relationships between the points of view held by mobilized and by mobilizing actors gives judgements as fine a degree of precision as one could wish for."[14] In a language closer to Nietzsche's and the neurophysiological insights that informed his epistemology: claiming that what is being observed exists as such only in relation to the means of its observation—that *how* something appears depends on the cells, organs, machines, technologies, representational, and semiotic systems that make it appear—is not to say that the observations that are made in this way cannot be falsified or deemed incorrect or would otherwise be indeterminate or nonbinding. On the contrary, Nietzsche recognizes that the linguistic conventions that are called truths are very much a "necessity" for human beings and for their ability to "exist socially" and to "make peace and ... banish from this world at least the most flagrant *bellum omni contra omnes*"; as "uniformly valid and binding designations invented for things," truths enable and secure our social existence (TL, 81). Nietzsche's famous epistemological skepticism, then, does not interfere, as Beiner insinuates, with one's ability to identify, for example, a politician's or reporter's lies, untruths, half-truths, fabrications, falsehoods, and misrepresentations. Instead, Nietzsche's conceptions of truth can help us understand better the post-truth attitudes of right-wing politics then and now: their lies target the existing social orders. Conversely, responding to such attacks by counting lies or with moral outrage only reinforces the social polarity the former hopes to expand. If Nietzsche's understanding of truth concerns the concept's ability to sustain power structures and values (including of decency, honesty, fairness, humaneness), then it might be advisable to put the latter (social relations and values rather than truths and lies) at the center of any assessment of these political movements.

Unfortunately, the liberal rejection of Nietzsche's philosophy often shares with its fascist appropriation the willingness to ignore the complexity of Nietzsche's arguments (and much of the extensive Nietzsche scholarship that deals with these complexities) in favor of isolating controversial statements mainly from his later writings that are then taken, as Beiner explicitly demands, at face value.[15] Such calls leave open which passages are to be taken at face value and which can be ignored. They certainly do not take at face value the many statements that Nietzsche makes wishing *not* to be read at face value, but rather creatively, as an "aphorism, properly stamped and molded, has not been 'deciphered' when it has simply been read, rather one has then to begin its *exegesis,* for which is required an art of exegesis" (GM, Preface 8). The demand to read Nietzsche at face value also ignores Nietzsche's extensive use of literary techniques (not just in *Zarathustra*), his reflections on the proximity of truths and metaphors, and that, in their aphoristic form, his writings constantly put their own assertions into new, different, and often contentious contexts. This is not to deny that some of Nietzsche's rhetorical choices are indefensible when viewed through the historical lens created by the humanitarian catastrophes of the twentieth century, and in particular by Nazi Germany. But this very fact should be taken as a warning rather than a license to ignore the intricacies of Nietzsche's thought.

A more complex image of Nietzsche's political thought has emerged in recent years among Nietzsche scholars who examine closely how his political stances, including his critique of democracy, might contribute to the development of "a revitalized and agonistic conception of democracy."[16] Siemens and Roodt's 2008 book includes several subsections dedicated to discussing Nietzsche's stance toward democracy. Scholars such as William E. Connolly, David Owen, Keith Ansell-Pearson, Alan Schrift, Lawrence Hatab, and Christa Davis Acampora (along with Siemens and Roodt themselves) draw on Nietzsche's work in support of democratic and liberal values. Hugo Drochon singles out in particular Owen, Schrift, and Hatab as scholars who "have seized upon Nietzsche's alleged decentring of the human being as a means of revitalising (American) democracy on a radicalised, postmodern basis, moving away from a conception of democracy too stuck, in their minds, in a religious and naturalistic vision of man now considered obsolete."[17] This group of scholars enlists Nietzsche in a tradition of political thought that has replaced consensus-driven political models with models that affirm, if not antagonism, then agonism as an indispensable basis of the political.[18] Drochon lists some of the different ways in which these scholars make Nietzsche "congenial to democracy": Connolly's idea of "agonistic respect"; Owen's concept of "agonistic deliberation," with a view to "ennobling democracy"; Hatab's "adversarial system"; or even Hannah Arendt's "robust public sphere," which "has also been linked to Nietzsche's idea of the agon."[19] Along similar lines, Donovan Miyasaki contends that, while Nietzsche is not a liberal thinker in the traditional sense, "we can find resources in his work to defend a noble egalitarianism: a unique, non-liberal form of political egalitarianism that is independent of classical liberal views from Locke to Rawls about essential human equality of worth and right."[20] Miyasaki adds that Nietzsche's promotion of "equality in the form of proportional, oppositional resistance . . . protects and promotes a diversity of human types and values" (168) and that, "consequently, a Nietzschean form of noble egalitarianism, far from leading to assimilation, would provide the only strong foundation for lasting, stable, and happy pluralistic societies" (169).

In sum, the existence of such starkly opposing views suggests that there is more at stake than that Nietzsche's philosophy "does not fit conveniently into political categories."[21] For one, poststructuralist interpretations of Nietzsche's philosophy have long argued that dominant dichotomies such as those between liberalism and fascism must themselves be rethought. Critical posthumanism can build on this tradition. As chapter 1 noted discussing Michel Foucault's distinction between humanism and the critical trajectory of the Enlightenment, political ideologies as varied as nineteenth-century liberalism, national socialism, and Stalinism draw on humanist themes in support of their agendas. Foucault's list makes clear how easily humanist ideals can be invoked and used to justify utterly inhumane actions and political modes of discrimination. If Foucault was correct, we should also not ignore the possibility that quite diverse political ideologies have more in common than they would like to admit. Specifically with regard to the liberal critique of Nietzsche's supposed fascist tendencies, it might be helpful (albeit not new) to consider the possibility that fascism is not the opposite of representative democracy, but its permanent dark "shadow," as Jan-Werner Müller substantiated recently with reference specifically to the resurgence of contemporary right-wing populist parties.[22] Furthermore, we can induce from Foucault's observation that posthumanism can offer an alternative perspective on traditional democratic–fascist, left–right, and similar political dichotomies by identifying and reflecting critically on the problematic humanist conceits that continue to inform different political ideologies and the ethos of some strands of contemporary posthumanism.

Beyond Biopolitics

Considering Nietzsche's influence on Foucault, it is perhaps not surprising that the first discussion of Nietzsche's political thought that recognizes explicitly his position as posthumanist and separates it from its far-right appropriation is indebted to Foucault. In his 2004 *Bios,* Roberto Esposito offers a far-reaching engagement of Nietzsche's writings and Nazism. He associates Nietzsche's political observations specifically with a posthumanist perspective, even entitling the subchapter that focuses most narrowly on Nietzsche "The Posthuman." Esposito credits Nietzsche for being the first thinker who understood that, "in the centuries to come the political terrain of comparison and battle will be the one relative to redefining the human species in a scenario of progressive displacement of its borders with respect to what is not human, which is to say, on the one hand to the animal and on the other to the inorganic" (83). Esposito reads Nietzsche's ethopolitical writings neither as promoting far-right politics nor as invested with the potential to enhance democracy, but rather as an astute analysis of what he identifies as the immunitary paradigm—in a nutshell, a rationality that justifies the negation of life for the purpose of protecting life that is central to the Western tradition of political thought. According to Esposito, "modernity makes of individual self-preservation the presupposition of all other political categories" (9). This paradigm is shared by modern prodemocracy as well as antidemocracy political parties. Esposito views Nietzsche as pivotal in the history of "immunitary *dispositifs*" because, in Nietzsche's later writings, the "entire immunitary semantic now seems to be rebutted, or perhaps better, to be reinterpreted in a perspective that simultaneously strengthens and overturns it, that confirms it and deconstructs it" (104–5).

With regard to the logic of immunization, Esposito argues that Nazism is not a negation of this paradigm (the protection of life), but a dangerous amplification of its inherent goal. Rather than identify and target perceived threats to the social body, in Nazism "the negative ... will be produced in growing quantities according to a thanatopolitical dialectic that is bound to condition the strengthening of life vis-à-vis the ever more extensive realization of death" (9). Esposito credits Nietzsche for having been "destined on the one side to anticipate at least on the theoretical level, the destructive and self-destructive slippage of twentieth-century biocracy, and on the other the prefiguration of the lines of an affirmative biopolitics that has yet to come" (10). Esposito also addresses the rhetorical missteps of the later Nietzsche,[23] contending that, while, "with Nietzsche, the category of immunization has already been completely elaborated" (47), "the more Nietzsche is determined to fight the immunitary syndrome, the more he falls into the semantics of infection and contamination" (96). Esposito thus takes Nietzsche's rhetoric as a sign for the dangerousness of the political territory he entered. His point is not to condemn or excuse Nietzsche, but to explore what Nietzsche has to say about this terrain.

Considering the prominence of themes of health, integrity, exceptionality, or perfection especially in Nietzsche's later writings, Esposito's biopolitical reading of Nietzsche and its focus on the immunitary *dispositifs* is certainly warranted and productive. Nevertheless, approaching the question of the ethics and politics of Nietzsche's posthumanism exclusively from a biopolitical perspective limits quite considerably the scope of one's inquiry. As Cary Wolfe has shown, the problem is that, while Esposito's affirmative biopolitical argument offers an alternative to the thanatopolitical thought associated especially with Giorgio Agamben's work, the grounding of "an 'affirmative' biopolitics in the material substrate of life itself as a given good"[24] resorts to a kind of neovitalism that comes at the expense of a closer examination of the sociopolitical, technological, and ecological constraints under which "life" acquires shape, meaning, and agency. Wolfe argues that what is needed instead "is what we might call an 'ecologization' of the biopolitical paradigm: an ecologization that drives us toward not a vitalism that wants to derive ethical or political imperatives from 'life' (in the form, let's say, of the 'biocentrism' familiar to us from the moment of 'Deep Ecology,' or a prima facie valuing of biodiversity in certain versions of environmental ethics) but rather a *denaturalized* understanding of the ecological paradigm that emphasizes form, time, and dynamic complexity . . . as the key constituents for thinking how biopolitics and bioart operate and signify."[25]

This book's extensive focus on Nietzsche's entomologically informed notion of sociality and on his reflections on technology has hoped to contribute to such an ecologization of biopolitical thought. We've also noted at various stages some of the limits of a biopolitical lens that conceives, as the basis of the self, a body that is thought, as Judith Butler puts it, as "a surface and a set of subterranean 'forces' that are, indeed, repressed and transmuted by a mechanism of cultural construction external to the body."[26] Doing so, we argued, this approach runs the risk of naturalizing the body (and life and nature itself), neglecting to understand it too as a cultural artifact that is defined, treated, and experienced differently by different cultures. Subsequently, the biopolitical approach does not think radically enough embeddedness and the constitutive force of human sociality and of technology. To reassess Nietzsche's ethopolitical thought along the lines of a "devitalizing ecologization of the biopolitical paradigm"[27] (as Erich Hörl describes Wolfe's critique of Esposito), this chapter will first return to the economic calculus on which the immunitary *dispositifs* Esposito identifies build and ask pragmatically what limits this calculus places on politics. At issue is the problem that the psycho-economic logic that continues to drive modern legal and political institutions (left and right) makes it difficult to address some of the most pressing political challenges society faces today (inequality, migration, climate change, war). These challenges, in a nutshell, cannot be solved by means of political, judicial, or even economic compensation or immunization.

This chapter will subsequently reflect further on some of the tensions between contemporary posthumanist articulations of community and the agonistic structure of Nietzsche's thought. Simply put, notions of community that are based on the recognition of "shared debt" (Vanessa Lemm) or on idealizations of "togetherness" (Rosi Braidotti) continue to put too much emphasis on the reasoning and moral rectitude of individuals. They are also ill-equipped to deal with disagreement. Because of these problems, the chapter will draw on Nietzsche's entomologically informed considerations of sociality and suggest replacing the venerable term "community" with the less revered term of the "swarm." The advantage of the latter concept is that it can better account for the contentious dynamics of social interactions and for the simultaneous dependence and independence of individuals vis-à-vis the social and technological systems in which they find themselves embedded. The concluding section will address the pragmatic implications of what this book has throughout identified as Nietzsche's "methodological posthumanism" and its radical displacement of "the human" into their social and technological environments. At issue is the question of how Nietzsche's critique of morality can contribute to a posthumanist ethos that avoids compensatory valuation regiments and more fully acknowledges the multiple ways (cognitive, social, and technological) humans are embedded. In all three regards, the following reflections on the ethics and politics of Nietzsche's posthumanism will venture beyond what could have been on Nietzsche's mind and instead relate and expand on his philosophy in ways that it deems relevant today. In line with recent scholarship that explores the potential usefulness of the agonistic structures of Nietzsche's thought for contemporary democratic concerns, the following considerations, then, should be viewed as Nietzschean rather than Nietzsche's.[28]

Resentment and the Politics of Calculation

It is more than tempting to project Nietzsche's famous concept of *ressentiment* onto fascist and modern populist movements. After all, aren't fear and hatred central to their raison d'être? Accusing the right of a politics of resentment has the added benefit that it makes it easy to "rescue" Nietzsche from his fascist appropriators, as he rejects so clearly "the poisonous eye of *ressentiment*" (GM I: 11) as a reactive and hence reprehensible force. Yet, the argument has serious flaws. For one, it can easily be turned against the accuser, as it invites the suspicion that it is driven by resentment itself. Moreover, it plays into the hands of what Jan-Werner Müller identifies as "a particular *moralistic imagination of politics*" typical for populist politics that pit a supposedly morally pure and unified fiction of the people against supposedly morally corrupt and inferior elites.[29] Resentment and anger might also have a valid role to play in democratic states, since resentment "is not merely concerned with individual self-esteem, wounded honor, or recognition, but also with the identification and protection of shared norms that regulate social and political relationships"[30]—an insight Michael Ure finds discussed already by the Scottish Enlightenment philosophers David Hume and Adam Smith. Furthermore, basing fascism on resentment reduces complex social and political dynamics to simple psychological attitudes. Rather than extend the category of *ressentiment* morally or psychologically, or distinguish between good and bad resentment, it is more circumspect to focus on the continued social and political relevance of the broader phenomenon Nietzsche places at the heart of the humanist cultivation process, what he identifies as its psycho-economic calculus. This focus reveals better how much of modern political discourse (left and right) and many modern political and legal institutions continue to revolve around a mode of thinking that is ill positioned to deal effectively with many of the most pressing challenges society faces today.

The previous chapter addressed Nietzsche's speculation that the "most primitive kind of cunning" might also have coincided with "the first appearance of human pride, man's sense of superiority over other animals," and that "perhaps the word 'man' (*manas*) expresses something of *this* first sensation of self-confidence: man designated himself as the being who measures values, who values and measures, as the 'calculating animal as such'" (GM II: 8). In the economic sphere, the ability to calculate forms the basis for the development of creditor–debtor relationships, which are subsequently transferred onto the legal rights of a person and become part of a body politics that constitutes itself by granting certain rights and protections to those who submit to its rule. While, over the course of centuries, the consolidation of power allows for more humane forms of "payment," including instances of mercy, such forms of humaneness extend the psycho-economic calculus and associated patriarchal gestures of a humanist tradition that allows individuals or groups to derive pleasure from putting themselves above others and, in the case of granting mercy, by putting themselves above the law. In an interesting reversal, Nietzsche suggests that a sense of justice is able to emerge only through the consolidation of power: challenging Eugen Dühring (who was also the target of Engel's influential polemic), he argues that "'just' and 'unjust' only start from the moment when a legal system is set up (and *not*, as Dühring says, from the moment when the injury is done)" (GM II: 11). If the primary function of the legal system is to control the violent propensities of injured parties seeking revenge by administering the law in a more measured way, justice remains linked to the archaic psychological dynamics that Nietzsche places at the beginning of the genealogy of penal codes, to a psycho-economic calculus that to this day seeks to create equivalencies by targeting individual subjects for punishment or reward. As we saw, justice thus helps promote rather than overcome the violence it claims to combat.

From a Nietzschean perspective, a posthumanist society worthy of this name would have to overcome this mode of thinking in equivalencies so central to the Western cultivation process and the institutions that sustain it. It would no longer seek to punish those who transgress its laws or bind people sociopolitically via a reward system built around, for example, party membership or a notion of citizenship that discriminates against those who do not share the same nationality. Indeed, considering Nietzsche's known conservative political leanings, it is only against the backdrop of his critique of these human, all too human modes of thinking that we can understand his (from today's perspective still highly) progressive positions vis-à-vis punishment or his rejection of nationalism, positions that are evident, for example, when he notes that "the purpose of punishment is to improve the one *who punishes*; that is the last resort of the apologists for punishment" (GS, 218), or when he claims: "No we do not love humanity; but on the other hand we are not nearly 'German' enough, in the sense in which the word 'German' is constantly used nowadays, to advocate nationalism and racial hatred and to be able to take pleasure in the national scabies of the heart and blood poisoning with which European peoples nowadays delimit and barricade themselves against each other as if with quarantines" (GS, 377).

Despite many efforts since Nietzsche's time to counter nationalism or to treat crime as a social rather than moral failure, it remains difficult to fathom the transition to a society that would no longer apply a compensatory logic. It is difficult to fathom not because humans are what they are, a critical posthumanist would note, but because of the technologies—the legal and political discourses and institutions—that have made this mode of thinking so durable. It is less difficult to recognize, however, the limited efficacy of legal and political institutions that are set up in this way. At issue is not just that the primitive psycho-economic calculus speaks to the underlying cruelty of humans who evolved within this calculus. Equally important is the question of its usefulness. For Nietzsche, this calculus had run its course and by the late nineteenth century was no longer able to address what he identified as the pressing issues of his time: broadly put, to promote life, to combat the threat of nihilism, to counter the drift toward mediocracy he associated with the advent of liberalism and mass culture, and so on. While the human-made threats to humanity of the twentieth and twenty-first centuries might be considered more existential, the political and judicial systems that ought to combat these dangers for the most part continue to be based on the same moral categorizations and economic calculus that Nietzsche put at the heart of the humanist tradition. The problem is that most of the major problems humanity faces today (from climate change, to economic inequality, to migration crises and medical pandemics or the ravages of war) cannot be solved or even be ameliorated by catering, simply put, to the psychological whims of individuals, or by rewarding or punishing certain constituents over others. In this regard, Nietzsche's narrative offers an explanation for why today much of politics and the media in their morally infused calls for action have become hindrances to addressing the pressing issues modern societies face. They incessantly replicate this mode of thinking and the institutions it supports rather than offer means and opportunities to coordinate the technological, institutional, and systemic changes needed to respond effectively to them.

Capitalism is a main culprit here. Not only is it the principal contributor to many of the most pressing dangers (climate change, inequality, exploitation, migration), it also continually instills economic calculation as a primary mode of thinking. Yet, following Nietzsche, who, as we noted in chapter 5, is quite critical of the values capitalism promotes, the problem is still more pervasive, as it concerns a primary aspect of the humanist cultivation process. Furthermore, reducing today's ills to capitalism alone runs the risk of reiterating a simple dichotomy between humans, on the one side, and society or its economic system, on the other, a dichotomy that, despite Marx's efforts to the contrary, often is cited in a way that ignores the social and technological embeddedness of humans. This is unfortunate, as the concept of embeddedness also allows us to recognize how individuals are complicit in constituting the systems in which they partake that affect their environments better than the oft-used dichotomy between "us" and society or capitalism does.

Insisting on an outside position of the self also downplays the extent to which technologies have changed and continue to change political and legal equations. In his 1990 essay on "Technology, Environment and Social Risk," Niklas Luhmann pinpoints the problem and contextualizes it historically, placing it well within the horizon of Nietzsche's time. In "the transition from the ethics of virtue to utilitarian ethics during the eighteenth century," the "apparatus of legally protected subjective rights presupposes a large array of possible actions which can serve the interests of the actor but will not have any harmful effects on others without their consent."[31] Modern technologies and the risks they present to the environment challenge this political rationale: the "danger produced by the risky decisions of others, which may eventually lead to large-scale catastrophes, can no longer be absorbed by contracts and payments and it therefore undermines a latent premise of our constitutional liberties."[32] While not necessarily new, the problem has become more apparent and more acute in today's globalized society where the economic and political interests of the most dominant actors are bound to contribute to the large-scale threats humanity faces as a whole and are prone to infringe on the rights of those least able to defend their interests (think migrants, *sans papiers,* the countries most exposed to the dangers of climate change, or the vulnerability of animals). In this regard, Nietzsche's warning about the dangers of Western humanism's underlying moral codes recognizes a systemic problem within the ideal of a civic state that is set up on the premise of a social contract.

From Posthuman Communities to Human Swarms

While the preceding chapters focused on the epistemological, social, and technological aspects of Nietzsche's posthumanism, each chapter also already reflected on some of their ethical and political implications. Chapter 2 argued that Nietzsche's epistemology advances a constructivist viewpoint that rejects ontological or materialist reductionisms and calls instead for a continued differentiation of linguistic, scientific, and technological processes that can more readily adapt to and engage the urgent ecological battles humanity faces. Chapters 3 and 4 examined how social formations understood as self-organizing aggregates question still-popular notions (and popular Nietzsche interpretations) that insist on the efficacy of willpower. We argued that the critical and constructive thrusts of Nietzsche's philosophy invite a diversity of viewpoints over the adherence to norms, foster creative responses over the protection of existing orders, promote affirmation over resentment, and thus help increase the flexibility of the emerging social, material, and technological networks they subtend. Chapters 5 and 6 focused on the question of how the hominization and cultivation effects of technology that Nietzsche identifies reconceive notions of self and subjectivity and contribute to the development of modes of thinking that are more attuned toward relationalities (social and across species), promoting the development of new sensibilities—and perhaps new policies—about our surroundings, the environment, and the embeddedness of humans therein. We argued that Nietzsche's recognition of the contingency of the Western cultivation processes opens up vistas onto a posthumanist future that would be constituted by alternative, less violent, less normative, more creative, self-reflective, and open cultivation processes driven by new interpretive schemas and communication possibilities.

Despite such apparent affinities, Nietzsche's philosophy is also offensive to a contemporary posthumanist ethos. His elitism and critique of "the herd" seem contrary to a "posthuman ethics [that] expresses a grounded form of accountability, based on a sense of collectivity and relationality, which results in a renewed claim to community and belonging by singular subjects."[33] Likewise, Nietzsche's well-known affirmations of power and modes of domination run counter to contemporary posthumanist sensibilities in which the critique of anthropocentrism is the foundation for a strong anti-authoritarian ethos and support for a situated pluralism attempts to help reduce social tensions and political conflict, as we saw Francesca Ferrando argue. The question is not whether it is possible to align Nietzsche with this ethos, but how the deconstructive force of his philosophy and of what this book outlined as his methodological posthumanism might comment on some of the contradictions inherent to this ethos. Furthermore, it remains to be seen whether Nietzsche's posthumanism can offer an alternative posthumanist outlook relevant for today.

Perhaps the most central point of ethical contention is the sense of community that posthumanists of different stripes like to envision as an alternative to the individualist conceits of the humanist tradition. Lemm's chapter "Posthumanism and Community of Life" goes to the heart of the matter. Despite her critique of Braidotti's "assemblage posthumanism" over the role Braidotti attributes to technology, Lemm and Braidotti agree a posthumanist ethics requires the creation of a new sense and forms of community. For Braidotti, community is situated and based on a new relationality. In her 2016 essay "Posthuman Critical Theory," Braidotti references Patricia MacCormack's work and subscribes to a "posthuman ethics [that] aims at enacting sustainable modes of relation with multiple human and nonhuman others that enhance one's ability to renew and expand the boundaries of what transversal and non-unitary subjects can become" (26). Such relations are guided not by "moral injunctions, but dynamic frames for an ongoing experiment with intensities that need to be enacted collectively, so as to produce effective cartographies of how much bodies can take, which is why I also call them: thresholds of sustainability." Thus, "posthuman ethics expresses a grounded form of accountability, based on a sense of collectivity and relationality, which results in a renewed claim to community and belonging by singular subjects." The problem is that such calls remain within the moral framework of a thinking that is the target of Nietzsche's critique. They continue to ground ethics in the rationality and moral capacity of individual subjects. While assemblage posthumanism adapts ethics to the complexities of contemporary societies and the modern pervasiveness of technology, its insistence on the centrality of the subject or subject formations still extends rather traditional humanist conceits.[34]

The biopolitics that Lemm detects in her reading of Nietzsche's *homo natura* figure similarly culminates in a model of life-affirming community that is thought to be posthumanist, inasmuch as it no longer revolves around the protection of subjects, but around affirmation. "As such, affirmative biopolitics is not the politics of the 'subject of immunitary protection,' as Wolfe ... argues, but a politics of community that is a politics of 'pure relation,' 'a relation without subjects.'"[35] In this context, Lemm suggests an understanding of justice that extends the economic calculus we saw Nietzsche place at the heart of the genealogy of morals to the point where it would sublimate itself:

> This striving for community comes hand in hand with striving for justice. Justice does not refer to a reciprocal, contractual relationship between equals based on an economy of exchange, which is in the best interest of the two parties. Instead, justice designates an asymmetrical relationship of gift-giving that is inherently a-economic and where what binds us to each other is the fact that we owe each other. We are bound to each other by an infinite responsibility to each other, an infinite debt that can never exhaust itself. It is in light of this debt that falls on us that we are equal. From the perspective of affirmative biopolitics, justice is based on gift-giving and expenditure. It is unliberal and anti-utilitarian.[36]

Based on the continued role that debt and compensation play here, the ethics Lemm proposes might perhaps better be called hyperhumanist or suprahumanist rather than posthumanist.

What distinguishes Lemm's from Braidotti's calls for community, however, is that Lemm's concept of community builds on the agonal structure of Nietzsche's thought. The argument, simply put, is that the continuity and success (in promoting excellence) of an agonistic society requires that the competing parties respect each other, a respect possible only if the competing parties are of comparable strength. Lemm pinpoints the logic behind Nietzsche's affirmation of competition when in the concluding paragraph of her book she adds:

> Nietzsche's belief that conflict and contest can only be productive and fruitful when power relations reflect an equilibrium—there can be no contest between the weak and the strong, but always only between forces that are more or less equal. Here equilibrium and equality are the starting point of productive conflict that results in the constitution of a juridical order that does not settle into a final and absolute political form.[37]

Lemm's analysis shows how Nietzsche's insistence on the agonal structure of social interaction can produce the kind of openness Braidotti calls for without having to rely on pleas for "togetherness," as we ought to answer the question "what 'we' are capable of becoming" in a "practical and pragmatic" way.[38] And yet, Lemm too grounds the hyperhumanist ethics she derives from Nietzsche in moral sentiments, in a shared sense of "infinite responsibility to each other" and of "infinite debt" that returns the argument to an ethical idealism anchored in the good intentions and reasonability of subjects and their willingness to support these ideals. From a critical posthumanist perspective, such arguments are problematic because their addressee is the rather familiar figure of the imaginative, rational, and well-meaning individual that was the hope of Enlightenment humanism all along. Behind this addressee hide in plain sight detached norms and moral expectations that sooner or later will elicit enforcement mechanisms in the form of punishment or reward. The argument is also pragmatically questionable. Historically, such appeals have hardly been very effective, or effective only when used to pit one party in the name of solidarity against opposing interests. The problem in universalizing the call for togetherness is that it adopts a humanist perspective that is prone to suppress opposition. Or, to recall the logic Esposito locates at the heart of modern politics: Sooner or later an ethical idealism will have to *immunize* itself against those who might not want to share these ideals, values, and perspectives. In short, such arguments are prone to turn against togetherness in the name of togetherness.

To avoid the reliance on anthropocentrically anchored moral codes and idealizations of togetherness, I want to expand on Lemm's recognition of the agonal structure of Nietzschean community formations and return to the entomological considerations of sociality we encountered in Nietzsche. That is, I want to adopt a more radically posthumanist perspective and replace the venerable term "community" with the admittedly much less revered term "swarm." Let's note right away that the point is not to reduce human beings and their dignity to that of insects; nor is it to ignore their diversity. On the contrary, by understanding social formations and their dynamics as swarm behavior, we can appreciate better the diverse "wills" of its members by no longer having to align and assess them along generalized moral codes, shared rationalities, or idealized posthumanist or hyperhumanist sensibilities. Instead, the swarm model makes it possible to take more seriously the emergence of values and the (temporary) success of particular perspectives from relational processes that are both enabled and limited by technological and environmental factors. Such a model is also better suited to account for the transitoriness and variability of individual and "public" opinions, intentions, and desires as individuals react differently to input, as ideas flourish and disappear, as motives develop and are abandoned again, as feedback loops build or cancel themselves out again, and so on.

Nietzsche's famous account of the genealogy of penal codes supports such a viewpoint. It recognizes the endogenous nature of evolutionary social processes where amplifications and cancellations create dynamic structures and drifts, as we would say today. In section 12 of the second essay—the de facto center of his *On the Genealogy of Morals* and the section where he explains most clearly how a genealogy differs from a historiography—Nietzsche returns to his critique of utilitarian and teleological interpretations of the history of morality (he mentions Herbert Spencer again), insisting that:

> There is no more important proposition for every sort of history than that which we arrive at only with great effort but which we really should reach,—namely that the origin of the emergence of a thing and its ultimate usefulness, its practical application and incorporation into a system of ends, are *toto coelo* separate; that anything in existence, having somehow come about, is continually interpreted anew, requisitioned anew, transformed and redirected to a new purpose by a

> power superior to it; that everything that occurs in the organic world consists of *overpowering, dominating,* and in their turn, overpowering and dominating consist of re-interpretation, adjustment, in the process of which their former "meaning" [*Sinn*] and "purpose" must necessarily be obscured or completely obliterated. (GM II: 12)

For the metaphysical idea of a telos directing the trajectory of social development, Nietzsche substitutes an agonistic model in which the continued struggle of forces underlies any appropriations of the meaning of a thing. In this context, Nietzsche separates the genealogy of the penal codes and modes of punishment from the meaning that is attributed to them: "The form is fluid, the 'meaning' [*Sinn*] even more so" (GM II: 12).[39] While this might sound "postmodernist,"[40] the point is that Nietzsche recognizes how both of these processes follow their own dynamics. Neither side fully controls the other, nor can they be reduced to the intentions or rationale of the individuals involved, as they themselves are informed by the historical processes within which they participate. If practice and meaning stand in no immediate causal relationship, then they must be thought to evolve in a process of self-modification in which changes on one side (practice) relate only indirectly to the changes noted (or ignored) on the other side (meaning), and vice versa. In other words, we are dealing with circular causalities that can be reduced neither to external causation nor to a form of interaction. They stand, as Thomas Fuchs put it with regard to enactive embodiment, in "a relation of implication or global-to-local encompassing."[41] Separating practice from meaning, doing from interpreting, Nietzsche not only questions the efficacy of linear causal explanations but also challenges us to think of the values (moral or utilitarian) attached to practices such as punishing not as a transcendental given or as justified by their effects, but as derived from ongoing social and interpretive struggles.

With regard to Nietzsche's fascist appropriation and his liberal critics, it is important to emphasize that the evolutionary model Nietzsche applies to the dynamics of social formations rejects any presumption of a teleology being at work that is central to Spencer's application of Darwin to the social realm. Thinking of social formations as evolving endogenously means to recognize their unpredictability and that they are not following an external rationale:

> So people think punishment has evolved for the purpose of punishing. But every purpose and use is just a *sign* that the will to power has achieved mastery over something less powerful, and has impressed upon it its own idea [*Sinn*] of a use function; and the whole history of a "thing," an organ, a tradition can to this extent be a continuous chain of signs, continually revealing new interpretations and adaptations, the causes of which need not be connected even amongst themselves, but rather sometimes just follow and replace one another at random. The "development" of a thing, a tradition, an organ is therefore certainly not its *progressus* towards a goal, still less is it a logical *progressus,* taking the shortest route with least expenditure of energy and cost,—instead it is a succession of more or less profound, more or less mutually independent processes of subjugation exacted on the thing, added to this the resistances encountered every time, the attempted transformations for the purpose of defence and reaction, and the results, too, of successful countermeasures. (GM II: 12)

By understanding Nietzsche's famous definition of genealogical processes as evolutionary along the lines of the interactive model of emergence he encountered in the entomological research of his time, it is possible to appreciate better the idea of constant "becoming" materially, institutionally, and in the realm of comprehension. From this perspective, there is no need to deny the "thing" agency. Instead, it becomes possible to recognize that the agency of "things" (here penal practices) is constituted and reconstituted continually by multiple independent processes of subjugation and the resistances that form around these attempted subjugations. While Nietzsche asserts a uniform "will to power" driving these struggles, this will, too, is situative: embedded in multiple competing relations and processes.[42] Any content, direction, or meaning of a thing is not defined by a singular will to power, but by the interactions between competing wills expressed in competing chains of interpretation. Nietzsche's affirmation of power, and of agonal and agonistic interactions thus recognizes and affirms the embeddedness of human beings, their situatedness, and with it the situatedness of their reasoning, as well as their continued becoming as they simultaneously contribute to and are swept up by the dynamics of these processes.

By conceiving of social and political dynamics along the model provided by swarms, we can also account better for a crucial factor in the evolution of social structures that is missing for the most part from biopolitical and assemblage posthumanisms: the recursive dynamics of interactions that take place in the social medium of communication. In swarms, dynamic structures and directionalities (becoming) develop not just in relation to outside (material or physiological) factors but also recursively, in relation to the swarm's own doings. Interacting with each other, the members constantly connect to, change, expand, build, or alter previous movements from which structures and drifts will emerge or that will modify their environments in ways that will alter subsequent interactions of swarm members with these environments. Eric Bonabeau, Marco Dorigo, and Guy Theraulaz call such intra-active effects "stigmergy."[43] In Nietzsche's language: wills constantly bump into, connect to, and compete with other expressions of will within the group and its environment. The recursive aspect of such interactive models is inherent to communication processes, as communication responds continually to other communication, leading to positive and negative feedback loops from which amplifications and (temporary) structures and drifts emerge. From this perspective, it is easier to explain the limited role the promotion of ideals and ideologies that call for unity play in the formation of public opinions for example: whatever guidance these ideals might offer, they can be expected to be swept up in interpretive chains and contentions that are the result of recursive amplification processes that may lead (to use a more popular vocabulary) to more or less polarization, a wider or more narrow spread of viewpoints, stronger or weaker forms of support for one or the other position, party, or political representative, and so on.

Needless to say, technologies, and in particular communication-media technologies, play a central role in the organization and amplification of such interactions. This is especially apparent today, in an age of algorithmic data mining where political parties follow the models set by companies like Alphabet Inc., Amazon, or Meta and correlate a vast amount of information about voters, their addresses, and their behavioral patterns as expressed in search histories and past votes (or purchases) to make suggestions, as N. Katherine Hayles recently observed, "that nudge you toward selections with methods that have, in a sense, pre-empted your decisions (creating a possible future for you based on the past/present enfolding)."[44] Such algorithmic steering of swarm behavior, though, does more than merely deprive individuals of decision-making by "pre-emptying" their choices, as Hayles rightfully laments; in exchange for the loss of autonomy (submitting to algorithmic selections), they simultaneously also give individuals greater autonomy in the form of more relevant choices. Furthermore, communication-media technologies offer venues for the formation of swarms that allow members to extend temporally and geographically the coordination of behavior far beyond what is possible among physically present neighbors. While the internet and today's social media add speed and increase the level of noise from which more varied structures (communication patterns) can emerge, the printing press and increased literacy rates in the eighteenth and nineteenth centuries had extended already in Nietzsche's time the possibility of swarms forming around shared interests and (situated) expertise well beyond the temporal and geographical boundaries experienced by oral cultures. Likewise, the possibility of politically and financially analyzing and exploiting the social dynamics communication-media technologies help facilitate existed already in the mass-media landscape of the late nineteenth century, although Nietzsche's time could match neither the algorithmic automatization nor the ubiquity of today's forms of data mining and metadata analysis.[45]

It bears repeating that abandoning moral calls that address "the subject" or posthuman subject formations as site and driver of social and political change does not mean to abandon expressions of individuality, agency, and difference. With Nietzsche, we can make the opposite claim: "With morality the individual is instructed to be a function of the herd and to ascribe value to himself only as a function" (GS, 116). John Richardson summarizes Nietzsche's counterintuitive stance on individuality: "The individual is ultimately a device of the common to improve itself."[46]

Appealing to individuals and their moral rectitude to support a shared sense of community and togetherness extends this humanist logic. Understanding "community" instead along the lines of the dynamics of swarms, as deriving from the contentious inter- and intra-actions of its members, makes it possible to better appreciate differences, including differences in perspective, values, and ideals, as well as the inherent limits narratives face that emphasize commonality and unity.

Lemm has argued that Nietzsche recognized two different forms of social organization, one geared toward the production of equals, and the other where "inclusion no longer means the equalization of the individual but the pluralization and diversification of the whole."[47] A diversified and pluralized whole multiplies the ways in which exceptionality can manifest itself. As Donovan Miyasaki puts it, "by promoting equality in the form of proportional, oppositional resistances, noble egalitarianism also protects and promotes a diversity of human types and values, since proportional power enables differing individuals and groups to resist domination and coercion by one another."[48] For Miyasaki, noble here means first and foremost that competition aims not for conquest, but "the preservation of proportional resistance and competition."[49] Nietzsche's "elitism" loses even more of its edge in light of his argument that agonistic relations and the exceptionality he hopes they promote are possible in all kinds of different contexts and pursuits. Nietzsche puts this perhaps most cogently in the fourth section of the preface to *Zarathustra,* which famously portrays "mankind as a rope fastened between animal and overman—a rope over an abyss." Conceptualizing human beings as "a bridge and not a purpose" allows Zarathustra to affirm in the same speech a slew of diverse life pursuits, from the "great despiser" to "the one who works and invents in order to build" to "the one who loves his virtues" and so on, pursuits that share with Zarathustra the willingness of "going under." What ought to go under or be overcome (Zarathustra uses both movements to announce the advent of the overhuman) is "yourself" (Z I: "On the Teachers of Virtue"), the "human being" (Z I: "On the Passions of Pleasure and Pain"; "On War and Warriors"), the "ego" (Z I: "On the Pale Criminal"), or even "life" (Z II: "On the Tarantulas"). Against the backdrop of Nietzsche's concept of agency as emerging with one's doings (see chapter 6), it should be noted that Zarathustra does not promote a thanatopolitical reaction in his affirmation of life. Rather, if put into a more prosaic language, Zarathustra here envisions a form of embeddedness in one's environment and one's doings where the doers experience themselves no longer as separated from their doings. We might call it an "authentic" existence, with "authenticity" defined not as the other side of what is perceived as artificial, contrived, fake, simulated, or otherwise staged, but as designating an identificatory relationship to their doings, where one no longer experiences oneself as separate from their doings— and thus no longer experiences themselves as a self.[50]

Nietzsche's entomologically informed model of social organization also offers an analytic tool to conceptualize social and political changes, and by extension the failure of attempts to anchor such changes in the moral convictions, ethical standards, or the reasonability of individuals. The point is not that reasonability or convictions would not matter, but that neither exists outside of the dynamics of the whole; nor can they fully control what evolves from the agonistic interactions between participants and their convictions, from wills clashing with other wills, as Nietzsche put it invoking specifically the interactions of ants. Nazi Germany offers the saddest example of this failure, as the society responsible for the worst human-made humanitarian crises in history was no stranger to humanist and enlightenment traditions and their moral standards. In a thought-provoking article on "The Hitler Swarm," Dirk Baecker provides a model for how swarm theory might conceptualize Germany's transition to Nazism and back to a functioning liberal democracy in a relatively short time. The point of such an analysis is not to serve as an apology for the perpetrators of crimes against humanity, but to understand better the failure of moral codes, reason, and other humanist conceits to prevent such human catastrophes from happening. In fact, unlike the popular idea of a mass that is thought to be "driven by external urge, and considered to be fallen victim to unknown forces," Baecker notes as an important difference that the concept of "a swarm presupposes endogenously motivated choices."[51] At issue is what Baecker identifies as the "emergent network effect any swarm exhibits. The swarm does not just draw on motivations to join it but produces motivations no individual would before have had knowledge of."[52] The model of the swarm thus allows us better to account for the complicity with a totalitarian regime of so many who did not support it before this regime took control in Germany or after its demise. It should also serve as a warning that the reliance on shared moral codes and reasonability are not a sufficient safeguard against the danger of a democratic society transforming into totalitarianism or fascism.

I noted in the conclusion to chapter 4 that, in his later writings, Nietzsche recognized that expressions of the will and of individualism are not opposed to the herd, but are rather part and parcel of the constitution of herds. "The will . . . is the herd," as Werner Hamacher puts it. Baecker spells out the underlying paradox of social formation more directly. Once swarm behavior is viewed as a form of social coordination through communication (not as guided genetically), it becomes apparent that social formations do not negate expressions of individual will or independence, but emerge from "the independence of individuals who *accept relations of dependence to increase their independence.*"[53] This paradox makes the notion of the swarm suitable "to help in the analysis of fascism, because under fascism individuals initiate a behaviour both emphasizing and denying their individuality."[54] Baecker continues to lay out in more detail than there is room for here the mechanisms of positive and negative feedback, of amplification and ambiguation, as well as the threats of violence that produced a swarm that simultaneously "was terrorized and terrorizing"[55] and that could turn "banal" individuals into conspirators and agents of mass murder.[56]

Toward a Pragmatic Posthumanist Ethics

In conclusion, I want to reflect on how a Nietzschean posthumanist ethics that rejects subject-centered moral valuations (not values altogether) may be applied today, to a society that is technologically advanced and complex to a degree that was surely beyond the purview of the nineteenth-century German philosopher. At issue is also the question of whether the methodological aspects of Nietzsche's posthumanism that I have emphasized throughout this book do not cater to a form of nihilism. Does the insistence on the primacy of the biological, social, and technological systems that humans embody and within which they find themselves embedded not signal a loss of agency that translates into political resignation that would only play into the hands of the cultural, economic, and political status quo? While the activist rhetoric of Nietzsche the "prophetic vitalist" might itself be seen as an attempt by the philosopher to counter this danger, I want to make the argument that the ecological aspects of Nietzsche's posthumanism this book has emphasized offer a pragmatically more feasible outlook. This means first and foremost that one resist projecting the complexities and intersections of modern society back onto a human or posthuman subject or subject formation. For, doing so levies an almost insurmountable burden, asking a person to deal not just with the "loss of humanist unity" and "the painful contradictions of the anthropocene moment" but also with the responsibility to serve as "building block for the next phase of becoming subjects together."[57] Rather than putting a posthumanist subject or subjectivity in charge of changing what Braidotti identifies as the multiplicity of material, technological, and discursive processes that define contemporary society, a more pragmatic approach will focus instead on the limits complex technological systems and social differentiation processes present regarding the possibilities of controlling or steering such systems. This is not to separate the private from the public along the lines in which Richard Rorty's neopragmatism wants "to treat the demands of self-creation and of human solidarity as equally valid, yet forever incommensurable."[58] Rather than put the modern self in a position of irony by running away from the critical lessons of Nietzsche (or poststructuralism) when it comes to the pursuit of human solidarity, a methodological posthumanism will strive to make these lessons productive for the continued reflection of the cognitive, social, and technological embeddedness of humans. Put bluntly, taking posthumanism's insight into the noncentrality of the human actor seriously means recognizing that the evolution of complex technological and social systems is neither reducible to nor controllable by the good or bad intentions of their human participants. Nor is it helpful to externalize and locate the perceived evils of modern society solely in its technologies and disciplinary regiments. This quandary has led methodological posthumanists like Latour and Luhmann, who embraced the noncentrality of "the human" perhaps most radically, to place human actors both on the inside and on the outside of social and technological differentiation processes. They are located inside inasmuch as they are elements of networks or systems, inasmuch as they participate in, contribute to, or resist their operations simultaneously as bodies, interlocutors, technicians, advocates, and so on. Yet, they are also on the outside of the dynamics of networks and systems, outside not in the sense of having oversight or control over that within which they are embedded, but the opposite: because individually they cannot steer or contain the evolution of such systems as a whole.

Acknowledging the noncentrality of individuals in social and technological processes does not mean that individuals, their intentions, or even their values are irrelevant. Rather, by acknowledging their noncentrality, attention can be returned to the situatedness and the agonistic structures within which individuals acquire limited agency and responsibility. In *Reassembling the Social,* when exploring his "political epistemology" (249–62), Latour makes a nonhierarchical understanding of society the basis for his critique of sociology as an academic science that positions itself outside of the social. He quips that "Nietzsche had traced the immortal portrait of the 'man of resentment,' by which he meant a Christian, but a critical sociologist would fit just as well" (252). His point is that "participants not scientists define what makes up the social work," which makes him return to the sociality of animals: "Just as a spider casts a web, economization is what is crafted by economists, socialization by sociology, psychologization by psychology, spatialization by geography" (257). In Latour's eyes, this should have made it impossible to conceive of sociology as a discipline that locates itself outside of society, as an objective observer.[59] It was only the nineteenth century "surprise," with the "emergence of masses, crowds, industries, etc." (258), that kept sociology alive, but that also contributed to the West's sense of superiority, which he argues needs to be rethought in a globalized world (see 262).

Luhmann's systems theory and his description of modern society as functionally differentiated offers perhaps a still more radical account for the exclusion of individuals as orchestrators or carriers of social processes. By placing not just subjects but also consciousness in the environment, rather than at the center of society (which he defines as made up not by individuals, but by the continued processing of communication and the dynamic structures and internal differentiations that emerge from communication), Luhmann offers a sophisticated model for the modern human's simultaneous embeddedness in multiple biological, cultural, social, and technological contexts or systems. In the short essay "Politische Steuerungsfähigkeit eines Gemeinwesens" (Political ability to steer a commonwealth), one of his last essays and one of the few places where Luhmann comes close to offering a pragmatic outlook, he suggests that a "cognitive openness" should replace the normative and emotional bonds that politics tends to emphasize. This would allow for a more realistic assessment of the state of affairs and offer new possibilities of "communicative behaviour," and Luhmann even wonders about the possibility of a "culture of provisional accommodation" that would "accept uncertainty as common ground and try to make determinations that can be modified should new circumstances or new insights arise" (61). As a consequence, the aspect of time would become more important for politics. Change would have to be expected, and priorities and values would have to be constantly adapted relative to such change. A politics geared in this way toward the future finds itself in a similar position as Nietzsche's posthumanist philosophy: it is oriented toward a future "post" that it can only announce but not envision yet. From this perspective, ethical standards and directives would have to become situative and open to change as their environments change. The recognition of change as a constant also necessitates the critique of moral codes that seek to establish stability, fixed grounds, immovable ideals, and the like. Werner Stegmaier notes as one of the aspects that Nietzsche's critique of morality shares with Luhmann's that both recognized how morality had sublimated itself to the point where it became morally necessary to warn about morality. Luhmann relates such warnings specifically to the risks a society takes that wants to "secure in the present a future it cannot know and thus is led to view morality itself as a risk" that could not only prevent the pursuit of certain possibilities to work toward securing society's future, but also infringe on the ability to "prepare for continued uncertainty and its continued need to change and adjust the guidance it might have to offer."[60]

Nietzsche shares with Luhmann and Latour the recognition that morality itself needs to be recognized, as Latour put it, as a "heterogeneous institution constituted from a multiplicity of events" that is itself "no more human than technology."[61] Viewed from this angle, Nietzsche's critique of morality can contribute to the development of an ethics that is no longer grounded in the moral capacity of individuals, notions of guilt, bad conscience, and related matters, but works instead toward a more pragmatic, situative assessment of what values should inform decisions in specific circumstances at a given moment. Based on its own insights into the way sociality and technologies challenge the coherence, autonomy, and centrality of "the human," a critical posthumanism, then, can and should promote context-specific ecological modes of thinking that are critical toward the forces that work against the increase of diversity and humaneness, but that also remain self-critical in the sense that they constantly seek to readjust their own codes and approaches in recognition of their multiple embeddedness in dynamic social and technological processes. "The work of mediation," as Latour puts it, "in its moral organization, requires ... the ceaseless circuit of concern, the penetrating return of scruple, the anxious reopening of the tombs in which automatisms have been heaped, the redeployment of means into partial aims and aims into partial ends."[62]

In her 2022 article on the "Ethics for Cognitive Assemblages," Hayles notes that the "mere fact that a space has to be carved out in which human autonomy can operate implicitly acknowledges the distributed agency, distributed decision-making, and distributed cognition characteristic of cognitive assemblages" (12). Hayles's essay offers an overview of the ethical quandaries the agential effects of technologies pose for moral codes that, to speak with Nietzsche, continue to be anchored in a psycho-economic calculus. Nietzsche's critique of morality can serve as a safeguard against reducing to questions of intentionality challenges that require a careful analysis of the complex distribution of agency among "participating entities, distributed decision-making at multiple levels among human and technical actors, and emergent meanings specific to the actors' internal and external milieus" (2). At the very least, Nietzsche's posthumanism foreshadows a lessening of the role moral judgments would play in the assessment of such complexities. As Nietzsche puts it in a note from 1887, a note that with its "ought" does itself not fully abandon a moral position: "One ought to reduce and contain step by step the realm of morality.... The territorial reduction of morality: a sign of progress" (KSA, 12:476).

How might such a reduction of the territory of morality look in practice? A recent article by Tamar Sharon on the ethical quandaries an affirmative biopolitics faces in the age of big data offers an example for how to approach in a more practical manner the biopolitical questions raised earlier. In "When Digital Health Meets Digital Capitalism," Sharon addresses the googlization of health research, and with it the many contradictions and ethical tensions high-tech companies pose and themselves have to confront, such as tensions between medical needs and privacy concerns, between economic imperatives and regulatory restrictions, between the promotion of innovations and their commercialization, and so on. Moral imperatives that focus on individual intentions have little to offer when it comes to balancing the various interests that at any time can work for or against improvements in the health of human beings in what are constantly changing biological, social, political, technological, and economic environments. While there is no issue with presupposing certain basic values and ethical standards, the challenge is how to balance the specific needs of interacting systems, and how to provide guidance under such circumstances while also remaining open to change as the technologies, economics, and political playing fields continue to change. Drawing on Luc Boltanski and Laurent Thévenot's "orders of worth," Sharon points specifically to Nietzsche, arguing that there is a need for "an ethics beyond morality (e.g. Nietzsche)" that allows for a more differentiated view of the complexities at hand and the embeddedness of the issues in question, recommending a situated "justification analysis" that would "help unpack conflicts by identifying the multiple values that are at stake in them that cannot always be reduced to solely civic vs. market orders of worth" (8). Guidance, Sharon suggests, could come from a "justification framework [in which] people find it important to justify what they do and that they will continually seek to justify their actions when confronted with criticism. This is a potent but still mostly untapped force in discussions on how to regulate the digital economy" (9). Put differently, in a posthumanist society, ethical guidance neither is anchored in an ideal of togetherness or community—or at least is not reduced to the propagation of such ideals—nor resorts to appeals to the moral capacity of individual subjects or posthumanist subject formations, but relies on a careful analysis of the situatedness of particular actors and the restrictions and limited choices they face at a particular point in time and in relation to multiple competing interests and dynamics. This means, in practice, to separate responsibilities, rather than try to unite them under the umbrella of shared values or ideals.[63] It also means to anticipate and accept resistance against shared values and ambitions and approach such resistance in a contextualized and strategic manner.

The preceding considerations of what this chapter identified as the ethical and political implications of Nietzsche's posthumanism, and more specifically of Nietzsche's critique of morality, might themselves be called moral. They certainly subscribe to a posthumanist ethos that supports the promotion of pluralism, equity, and human and animal dignity, as much as they see the pressing need to address global dangers such as climate change, the humanitarian crises caused by forced migration, the dangers and ravages of war, and global health crises. At issue are not the basic values or ideals that inform this outlook, but the modes of thinking within humanist and some posthumanist traditions that work against these goals. The critical thrust of Nietzsche's philosophy offers means to delineate some of these inherent contradictions. While the concerns raised by critical and methodological strands of posthumanism and by the close examination of the complexities of Nietzsche's philosophy might quickly lead to high levels of abstraction, they are not without practical implications, including for critical posthumanism itself. They entail at the very least a change in perspective, away from the articulation of shared moral values that rely on the good sense and intentions of individuals and toward a closer analysis of the structural complexities and the situatedness that foreclose but also open up possibilities of action for those involved. Regarding the particular situatedness of this book on Nietzsche's posthumanism: the hope has been that the dialogue it forged between contemporary posthumanism and Nietzsche's philosophy will offer perspective and new outlooks on both. The hope has likewise been that this study will thus contribute in its own limited way to the larger project of a critical posthumanism wishing to increase thought and communication possibilities that can better account for and address the complexities that emerge where human, all too human modes of thinking intersect with posthuman, all too posthuman predicaments.

Notes

Introduction

1. Soper, "Humanism in Posthumanism," 375.
2. Bennett, *Vibrant Matter*, ix.
3. With regard to Nietzsche's styles, Derrida famously expressed Nietzsche's loss of authorial control in the metaphor of the spider, arguing that Nietzsche might have been caught in his writings "much as a spider who finds he is unequal to the web he has spun" (*Spurs*, 101).
4. Robert C. Holub's 2018 study is the most recent to situate Nietzsche extensively in the "discursive universe of the late nineteenth century in Europe" (*Nietzsche in the Nineteenth Century*, 10). For a summary of Nietzsche's knowledge of the natural sciences, see Brobjer, "Nietzsche's Reading and Knowledge of Natural Science." Brobjer finds that about one hundred titles, or one-tenth of the books in Nietzsche's private library, belonged to the field of natural science.
5. Moore, "Introduction," 2.
6. Foucault, "What Is Enlightenment," 11.
7. Heidegger, *Basic Writings*, 233.
8. See Maturana and Varela, *Autopoiesis and Cognition*. See also Thompson and Varela, "Radical Embodiment," for a concise presentation of how the cognitive sciences can move past the mind–body and mind–world dichotomies by conceiving of "brain, body and environment . . . as mutually embedded systems rather than as internally and externally located with respect to one another" (423–24).
9. Morton, *Hyperobjects*, 22.
10. Hayles, "Cognitive Nonconscious and the New Materialisms," which is also chapter 3 in her *Unthought: The Power of the Cognitive Nonconscious*.
11. Lemm, *Nietzsche's Animal Philosophy*, 9.
12. Derrida and Beardsworth, "Nietzsche and the Machine," 31.
13. Lemm, *Homo Natura*, 83.
14. Ferrando, *Philosophical Posthumanism*, 151.
15. Braidotti, "Posthuman Critical Theory," 27.

1. Posthumanism and Its Nietzsches

1. Pepperell, *Posthuman Condition*, 1.
2. Nayar, *Posthumanism*, 2.
3. Meillassoux, *After Finitude*, 124.
4. Lemm, *Homo Natura*, 11.
5. Braidotti, "Theoretical Framework," 33.
6. Esposito, *Bios*, 83. It is under the subchapter heading "Posthuman" (in ch. 3, "Biopower and Biopotentiality") that Esposito offers the most detailed discussion of Nietzsche's ethopolitical stance.
7. The second chapter of Stefan Herbrechter's *Posthumanism* is titled "A Genealogy of Posthumanism." Herbrechter puts Nietzsche first, although he also recognizes that phenomena associated with posthumanism have a longer history (see Herbrechter and Callus, *Posthumanist Shakespeares*).
8. Bostrom argues that Nietzsche's "vision" of the overhuman does not align with transhumanism's "humanistic concerns for the welfare of all humans (and other sentient beings)" ("History of Posthumanist Thought," 4). While Nietzsche indeed questions liberal ideals we today intuitively associate with human welfare, it is disingenuous to claim that Nietzsche does not want to promote the welfare of humanity. Nietzsche questions, however, Western society's particular ideal of human welfare, an ideal that Zarathustra, with the figure of the "last human," boils down to the pursuit of comfort and petty pleasures. Bostrom's position in turn is opposed by Stefan Sorgner, who thinks that "significant similarities between the posthuman and the overhuman can be found on a fundamental level" ("Nietzsche, the Overman, and Transhumanism," 30). The debate about Nietzsche's transhumanism is continued in *JET* 21, no. 1, and was republished and extended in Yunus Tuncel's 2017 edited volume *Nietzsche and Transhumanism: Precursor or Enemy?* Ansell-Pearson's *Viroid Life* explored already in 1997 continuities and differences between Nietzsche and transhumanism, focusing especially on Nietzsche's philosophy of life. I will return to this debate in more detail in chapter 5.
9. Pippin, *Nietzsche, Psychology, and First Philosophy*, 5.
10. Luhmann, *Die Wissenschaft der Gesellschaft*, 112. For an extensive discussion of the many continuities between Luhmann and Nietzsche, despite Luhmann's limited acknowledgment of Nietzsche, see Stegmaier, *Orientierung im Nihilismus*.
11. For a more recent example of Nietzsche's significance for transhumanist speculation, see Fuller's 2020 *Nietzschean Meditations*.

12. Lemm reads Nietzsche through the lens of an "affirmative biopolitics" that "sees in the continuity between human and animal life a source of resistance to the project of dominating and controlling life-processes" (*Nietzsche's Animal Philosophy*, 9).

13. Ferrando, *Philosophical Posthumanism*, 151.

14. Wolfe, *What Is Posthumanism?*, xvi.

15. See Beiner's 2018 book *Dangerous Minds* for a troubling summary of leading right-wing figures appropriating Nietzsche for their nefarious political ambitions. I will return to the political implications of Nietzsche's posthumanism and to Beiner's critique of Nietzsche in chapter 7.

16. For a more differentiated view of the antecedents to posthumanism around 1800, see Landgraf, Trop, and Weatherby, *Posthumanism in the Age of Humanism*.

17. Nayar, *Posthumanism*, 2.

18. See Herbrechter and Callus, "Introduction—Shakespeare Ever After."

19. Herbrechter, *Posthumanism*, 17.

20. Braidotti, "Posthuman Critical Theory," 17.

21. Braidotti sees cartographies as essential conceptual tools that ought to be "both critical—of dominant visions of knowing subjects—and creative—by actualizing the virtual and realized insights and competences of marginalized subjects" ("A Theoretical Framework," 39).

22. Nayar, *Posthumanism*, 3.

23. Wolfe, *What Is Posthumanism?*, xvi.

24. Wolfe, xvii.

25. Herbrechter and Callus, "Introduction—Shakespeare Ever After," 6.

26. Hayles, *How We Became Posthuman*, 287.

27. Wolfe, *What Is Posthumanism?*, xviii.

28. Vita-More, "Tanshumanist Manifesto."

29. Herbrechter, *Posthumanism*, 32.

30. Braidotti, *The Posthuman*, 12.

31. Wolfe, *What Is Posthumanism?*, xxvi.

32. Heidegger, *Platons Lehre*, 49.

33. See Heidegger, *Basic Writings*, 224. All quotes from the *Letter on Humanism* are taken from Heidegger, *Basic Writings*, with page numbers subsequently being referenced in parenthesis in the text.

34. Heidegger, *Platons Lehre*, 50. I am bracketing Heidegger's critique of Nietzsche for failing to think fully the essence of being outside the metaphysical tradition, and hence for not fully leaving behind a metaphysical mode of

thinking (something Nietzsche, of course, acknowledges at various points), for a discussion of this critique would lead too deeply into Heidegger's ontology, and thus too far astray from a discussion of Nietzsche's posthumanism.

35. Adorno's book analyzes inherent contradictions in Heidegger's (and his followers') "jargon of authenticity" and chastises as elitism a language he views as a "trademark of societalized chosenness, noble and homey" (*Jargon of Authenticity*, 5–6). It should also be noted that Heidegger does not use the *Letter on Humanism* to reflect on his personal involvement or on possible affinities of his own thought with Nazi ideology. The only reflective moment we find in the *Letter on Humanism* is a defense of his turn against humanism in his 1929 magnum opus *Being and Time*.

36. Nancy, *Banality of Heidegger*, 41.
37. Wolfe, *What Is Posthumanism?*, xxvi.
38. Colebrook, *Death of the PostHuman*, 109.
39. Bennett, *Vibrant Matter*, 12.
40. In the *Letter on Humanism,* Heidegger subsequently closes the circle that dovetails humanism and metaphysics by arguing that any definition of the essence of humans that does not inquire about the truth of Being is metaphysical, and hence that "all metaphysics . . . is 'humanistic.'" (*Basic Writings,* 226).
41. Graham, "Nietzsche Gets a Modem," 72.
42. Sharon's mediated posthumanism subscribes to the nonessentializing tendencies of radical and methodological posthumanism, but takes a more pragmatic stance, inquiring about the transformative effects modern biotechnologies have on subjectivity.
43. Sharon, *Human Nature,* 5. Sharon includes transhumanist theorists like Nick Bostrom, James Hughes, Ray Kurzweil, Hans Moravec, and Julian Savulescu in this group and adds other liberal approaches to new biotechnologies she finds in the works of Nicholas Agar, Allen Buchanan, and John Harris.
44. Sharon adds Neil Badmington, Anne Balsamo, Rosi Braidotti, Elaine L. Graham, Chris Hables Gray, Donna Haraway, N. Katherine Hayles, Allucquère Rosanne Stone, and Joanna Zylinska to the list of radical posthumanists.
45. Sharon, *Human Nature,* 6.
46. Braidotti's *The Posthuman* offers an example where the rift I am noting here does not separate different strands of posthumanism, but runs right through the work of one of its most visible theoreticians. Braidotti defines posthumanism as the historical moment that marks the end of the opposition between humanism and antihumanism and traces a different discursive framework. But the discursive framework she uses regularly returns to an anthropocentric viewpoint

that reaffirms a normative concept of subjectivity, as is apparent in her frequent use of first-person pronouns (which serve not just as contingency markers) and in her insistence on the need for "at least *some* subject position" (102) with the ability to "autopoietically self-style" in ways that accord with "what we humans truly yearn for" (136). Even if the object of this yearning is subsequently qualified as a desire "to disappear by merging into this generative flow of becoming, the precondition for which is the loss, disappearance and disruption of the atomized, individual self" (136), the argument returns us to a humanist view that asserts an inherent human nature (as becoming) and reaffirms a normative notion of subjectivity. Chapter 6 will return to this tension, which remains present also in Braidotti's more recent writings.

47. Sharon, *Human Nature,* 121n7.
48. Sharon, 5.
49. Sharon, 6.
50. Sharon, 9.
51. See Hayles, "Cognitive Nonconscious" and *Unthought.*

2. Posthumanist Epistemology

1. Butler, "Performative Acts," 520.
2. Braidotti, "Posthuman Critical Theory," 14.
3. On Nietzsche's early dissertation project and his readings of natural scientists, see Solies, "Naturwissenschaften als Aufklärung?," 249.
4. Emden, *Nietzsche on Language,* 83. Already in the 1970s, Friedrich Kittler noted how, during the same epoch that physiologists (Kittler mentions Helmholtz and Gustav Fechner) calculated sensory thresholds, Nietzsche described the sensory production of differences and intensities (see "Nietzsche (1844–1900)"). For an excellent summary of the psychophysiological research with which Nietzsche was familiar, focusing in particular on the "frameworks of evolutionism and monism advocated by Ernst Haeckel," see Schloegel and Schmidgen, "General Physiology, Experimental Psychology, and Evolutionism," 614. See also Holub, *Nietzsche in the Nineteenth Century,* on the significance for Nietzsche's later thinking of Wilhelm Roux's *The Struggle of the Parts in the Organism* (1881), William Henry Rolph's *Biological Problems* (1884), and Carl von Nägeli's *A Mechanical-Physiological Theory of Organic Evolution* (1884).
5. On Müller's research methods and his significance for nineteenth-century physiology, see Otis, *Müller's Lab.*
6. Müller, *Handbuch der Physiologie,* 688.

7. Müller, *Über die phantastischen Gesichtserscheinungen*, 5.

8. Jakob von Uexküll's remarkable book *The Meaning of Life* (written in 1942, published in 1947) purports to find the answer to this audacious question in Müller's 1824 inaugural lecture "On Physiology's Need for a Philosophical Observation of Nature" (reprinted as part of *Zur vergleichenden Physiologie*). Uexküll warns at the very beginning of his book against confusing the concept of specific energies popular in Müller's time with the now dominant understanding of energy as a physical (i.e., constant) quantity. At stake is not a form of energy that would contradict the second law of thermodynamics, but a concept that references the formation of specific biological structures, processes, and entities (such as cells) that are embedded in, and yet relate autonomously to, their (organ-specific) environments. Uexküll suggests it was Goethe's *Metamorphosis of Plants* that offered Müller "immediate insight into the form-giving activity of life, which does not need causality" (*Sinn des Lebens*, 66).

9. Müller, *Handbuch der Physiologie*, 272.

10. Crary, *Techniques of the Observer*, 92.

11. Lotze, *Allgemeine Pathologie*, 150 (trans. Woodward, "Hermann Lotze's Critique," 152).

12. Woodward, "Hermann Lotze's Critique," 155–56.

13. Woodward, 156.

14. See Evan Thompson and Francisco J. Varela for a concise presentation of how the cognitive sciences can move past the mind–body and mind–world dichotomies by conceiving of "brain, body and environment . . . as mutually embedded systems rather than as internally and externally located with respect to one another" ("Radical Embodiment," 423–24). In his recent article, Thomas Fuchs expands on Thompson and Varela's model, describing as the basis of the embodied mind "two interactive or feedback cycles, (a) Cycles of organismic self-regulation, engendering a basic bodily sense of self; and (b) Cycles of sensorimotor coupling between organism and environment, implying an 'ecological self'" ("Circularity of the Embodied Mind," 3). By accepting circular (i.e., organistic in Müller's sense) causalities, an enactive approach makes it possible to understand how "the phenomenology of bodily being in the world corresponds to the ecology of the organism in relation to its environment. Lived body and physical body are both complementary aspects of the same life process that connects the living subject and the world, or the brain, body, and environment in circular interactions" (6).

15. Engler-Coldren, Knapp, and Less, "Embodied Cognition," 417.

16. Michael Heidelberger finds differences primarily in degree, arguing

that Helmholtz's understanding is simultaneously more physical and more spiritual-idealistic than Müller's ("Beziehungen," 44). Theodor Leiber offers a more detailed look at the philosophical differences between student and teacher (*Vom mechanistischen Weltbild zur Selbstorganisation des Lebens,* 411).

17. Emden states that Nietzsche read Helmholtz's book in the early 1870s (*Nietzsche on Language,* 95). Considering his interests in music and physiology, it is likely that Nietzsche was familiar with Helmholtz's book in the late 1860s already. It would help explain why Nietzsche put the title of Helmholtz's 1866 book under Müller's name. In either case, it is safe to assume that Nietzsche was familiar with the principle of specific nerve energies when he wrote *On Truth and Lies in a Nonmoral Sense* in 1872/1873.

18. Helmholtz, *Die Lehre von der Tonempfindung,* 13.

19. Nietzsche's epistemology signals what Günther Abel describes as the transition from a thing model to an event or process model in relation to the problem of consciousness ("Bewusstsein—Sprache—Natur," 11).

20. See Müller, *Zur vergleichenden Physiologie,* xii.

21. Nietzsche continues: "If truth alone had been the deciding factor in the genesis of language, and if the standpoint of certainty had been decisive for designations, then how could we still dare to say 'the stone is hard,' as if 'hard' were something otherwise familiar to us, and not merely a totally subjective stimulation!" (TL, 80–81).

22. Crary, referencing Schnädelbach, notes that Schopenhauer already presents us a "physiological reinterpretation of the Kantian critique of reason" in which "the notion of correspondence between subject and object disappears" (*Techniques of the Observer,* 75). My interpretation of Nietzsche as offering a radicalized interpretation of Schopenhauer is diametrically opposed to Maudemarie Clark's, who reads the passage as evidence that "Nietzsche apparently thinks that common sense affirms the independent existence of the external world" (*Nietzsche on Truth and Philosophy,* 81). Clark subsequently distinguishes between "things themselves" and the thing-in-itself, claiming that, when Nietzsche speaks of the former, he "refers to the world considered as having *existence* independently of perception or representation" (82). Clark's account is puzzling not only because it neglects the physiological argument made in the essay, and for its stern reliance on a simplistic bivalent logic in the face of Nietzsche's more complex, often paradoxical thought figures, but also because it never reflects on why the assumption of a super-sensory "existence" should elude the metaphoric and anthropocentric limits Nietzsche puts on language.

23. For Schopenhauer, "what the eye, the ear, or the hand feels, is not

perception; it is merely its data. By the understanding passing from the effect to the cause, the world first appears as perception extended in space, varying in respect to form, persistent through all time in respect of matter; for the understanding unites space and time in the idea of matter, that is, causal action [*Wirksamkeit*]" (*World as Will and Idea*, 36).

24. See Emden, *Nietzsche on Language*, 96. Likewise, the "physiological description of sensory perception as a translation process, originally formulated by Büchner, Lotze, and Funke, appears fairly commonly in nineteenth-century works" (105), including in the writings of Helmholtz, Fechner, Eduard von Hartmann, and others. Emden further notes that Friedrich Albrecht Lange, in his 1866 *Geschichte des Materialismus,* also already connects the nervous system to language.

25. Chladni is mentioned in Helmholtz's *Die Lehre von der Tonempfindung* (*On the Sensation of Tone*). Helmholtz describes how a plate covered with sand can be used to visualize frequencies and credits Chladni for having discovered a number of interesting phenomena of such mappings of sound (70).

26. Fuchs, "Circularity of Embodied Mind," 4. Fuchs continues: "If that is the case, then primary self-awareness can no longer be localized anywhere in the brain; rather, it is the integral manifestation of the brain–body system, or of the overarching process of life, which encompasses the whole organism. The same applies to emotions: as resonant loops between the brain, body, and environment, they are no longer the brain's representations of the body's activity, as Damasio puts it, but rather the feelings of the body itself vis-à-vis a certain situation" (4).

27. See Tobias Wilke, "At the Intersection of Nervous System and Soul," on how the turn from visual to auditory conceptions of cognition can be observed in many of the physiological and psychological writings of the late nineteenth century and well into the twentieth century, e.g., in the writings of Friedrich Theodor Vischer, Theodor Lipps, or Johannes Volkelt.

28. Müller, *Über die phantastischen Gesichtserscheinungen,* v. In this work, Müller investigates "fantasy" in the form of phantasmagoric and other visions along with his neurophysiological studies, because it is here that the astute observer can catch the visual sensory substance at work independent of external stimuli. For a more detailed reading of the treatise in relation to posthumanist epistemological contentions, see Landgraf, "Embodied Phantasy."

29. See Müller, *Über die phantastischen Gesichtserscheinungen,* 4–5.

30. Müller, 5.

31. With these observations, Müller explicitly challenges traditional teachings on optics, which assume light either as entering the eye or as immanent to

the eye. Müller points toward an "obvious" physiological contradiction in this viewpoint. "How," Müller asks, "even if there was something outside that was luminous itself, could this objective light reaching the subjective areas also be sensed subjectively as illuminate? . . . It would be equally inappropriate to say that the [sensory substances] would themselves sound, be heated, or carry a certain taste" (*Über die phantastischen Gesichtserscheinungen,* 8). Müller instead understands light as sense energy (*Sinnesenergie*) and what the visual sensory substance produces as self-illumination (*Selbstleuchten*). See *Zur vergleichenden Physiologie,* 395.

32. Today, memes might be understood along these lines, as organicized concepts; though, with memes, we usually think of their virus-like replication within the social media sphere, rather than of their epistemological quality, their ability to create their own realities.

33. For a contextualized reading of Nietzsche's use of the hinge metaphor in *On Truth and Lies* and his attempt to counter language's tendency to eradicate differences, see Holland "Angeln, Blatt, Constellation."

34. In *On Truth and Lies,* Nietzsche uses a building metaphor to describe the creation of the conceptual world as "the piling up [of] an infinitely complicated dome of concepts upon an unstable foundation, and, as it were, on running water" (TL, 85).

35. Lange, *Geschichte des Materialismus,* 727.

36. See Crawford, *Beginnings of Nietzsche's Theory of Language,* 68. Crawford shows how Lange is referring to Müller's work on the physiology of sight, but also critiques Müller, who in Lange's assessment remains within the theory of projection of images as he remained entangled in the notions of subject and object (72). Lange concludes the chapter on "Physiology of the Sensory Organs and the World of Representation" by returning to the figure of the circle, arguing that "the explanation of all that is psychological in terms of brain and nerve mechanisms is the safest way to reach the insight that it is here where the arch of our cognizing [*der Bogen unseres Erkennens*] closes itself without ever touching on what the mind [*Geist*] is in itself" (*Geschichte des Materialismus,* 735).

37. Lemm, "Shedding New Light," 352. Lemm is summarizing a main concern of "naturalist readings of Nietzsche" as described by Christopher Janaway and Simon Robertson, *Nietzsche, Naturalism and Normativity,* 5.

38. Bennett, *Vibrant Matter,* 5.

39. Ferrando, *Philosophical Posthumanism,* 6 and 185. Ferrando engages neither the neurophysiological basis of Nietzsche's epistemology nor the underlying paradox.

40. Ferrando, 12.

41. The importance of these considerations of causality, which are based on neurophysiological findings of his time, is evident in the fact that Nietzsche reflects on them prominently still in one of his last works, *Twilights of the Idols*, where, in the chapter "The Four Great Errors," he returns to the false attribution of causes in dreams when "a particular sensation, for instance, a sensation due to a distant cannon shot, has a cause imputed to it afterwards (often a whole little novel in which precisely the dreamer is the protagonist).... The cannon shot shows up in a *causal* way, and time seems to flow backwards.... What has happened? The representations *generated* by a certain state of affairs were misunderstood as the cause of this state of affairs" (TI, 32–33).

42. The passage anticipates what, in GM I: 13, Nietzsche identifies as a seduction of language: the separation between doer and deed, a precondition for morality, and thus a major accomplishment in the history of civilization. The passage in HH shows how important the conclusions are that he draws from nineteenth-century neurophysiology including for his critique of morality. Chapter 6 will expand on the significance of Nietzsche's observation regarding language's separation of doer from deed.

43. Heidelberger summarizes the main differences between Müller and Descartes and between Müller and Kant. For Müller, it is the material living body, not the spirit as in Descartes, that becomes aware of itself as subject. The certainty of the cogito for Müller is the certainty of the body sensing itself. And unlike for Kant, Heidelberger adds, the subject is for Müller not transcendental ("Beziehungen," 43).

44. See Wellbery, "Die Strategie," 159.

45. See Colebrook, *Death of the PostHuman*, 109.

46. Donna Haraway has recently questioned the concept of autopoiesis, arguing that "Nothing makes itself; nothing is really autopoietic or self-organizing" (*Staying with the Trouble*, 58). Haraway proposes instead the use of the term "sympoiesis," to highlight the embeddedness and connectedness between systems and the material environment within which they emerge and operate. It would seem to me that Haraway's suggestion does not contradict, but highlights a central point of second-order systems theory: systems operate always already in a system-specific environment; i.e., the environment is indeed coconstitutive of the system and its operations.

47. Though Luhmann rarely mentions Nietzsche—and when he does, without much reverence—he lists Nietzsche along with Heidegger and Derrida among those thinkers who realized the paradoxical nature of observation (*Die*

Wissenschaft der Gesellschaft, 94). Stegmaier expands on the continuities between Nietzsche's perspectivism and Luhmann's constructivism, on how both thinkers replace the idea of reality as an "outside" with the idea that reality is the product of reflexively and recursively operating modes of observation, in chapter 2 of *Orientierung im Nihilismus*.

48. Luhmann, *Die Wissenschaft der Gesellschaft*, 718–19.
49. Luhmann, 718–19.
50. Luhmann, 719.
51. Engler-Coldren, Knapp, and Less, "Embodied Cognition," 417.
52. Hayles, *How We Became Posthuman*, 4.
53. Hayles, 49.
54. Sharon, *Human Nature*, 30.
55. Nayar, *Posthumanism*, 2. For Nayar, "Posthumanism is about the embedding of embodied systems in environments where the system evolves with other entities, organic and inorganic, in the environment in a mutually sustaining relationship" (51).
56. Hayles, *How We Became Posthuman*, 5
57. Wolfe, *What Is Posthumanism?*, xv. Interestingly, Wolfe includes Hayles in his critique of the transhumanist tradition, suggesting that, while Hayles is critical of transhumanism, she nevertheless appears to associate the posthuman with a "triumphant disembodiment" (xv).
58. Meillassoux, *After Finitude*, 25. Meillassoux also here calls the body the "'retro-transcendental' condition for the subject of knowledge."
59. Meillassoux, 26.
60. Meillassoux, 115. As Leif Weatherby points out, "Meillassoux defines his project as the construction of a non-metaphysical speculation, one founded in a mathematical language that Meillassoux has not provided" ("Farewell to Ontology," 155). To date, Peter Wolfendale's *Object-Oriented Philosophy: The Noumenon's New Clothes* offers the most comprehensive critique of OOO and speculative realism.
61. Hayles, "Cognitive Nonconscious," 181.
62. Hayles, 181.
63. Meillassoux, *After Finitude*, 124.
64. Meillassoux, 37. Meillassoux continues here: "A metaphysics of this type may select from among various forms of subjectivity, but it is invariably characterized by the fact that it hypostatizes some mental, sentient, or vital term: representation in the Leibnizian monad; Schelling's Nature, or the objective subject-object; Hegelian Mind; Schopenhauer's Will; the Will (or Wills) to

Power in Nietzsche; perception loaded with memory in Bergson; Deleuze's Life, etc." In this chapter I hoped to substantiate that Nietzsche's epistemology neither is based on a new form of subjectivity nor belongs to what Meillassoux in this context (and with specific reference to Nietzsche and Deleuze) calls "antirationalist doctrines of life and will" (37).

    65. Morton, *Hyperobjects*, 22.
    66. Clarke, *Gaian Systems*, 14.
    67. Müller, *Über die phantastischen Gesichtserscheinungen*, 106.

3. Insect Sociality

    1. See Latour, "On Recalling ANT."
    2. Latour, *Reassembling the Social*, 9.
    3. Deborah Gordon offers accessible accounts of the diversity of ant behavior and their marvelous accomplishments. She advocates the use of cybernetics and network theory to understand how, in ant colonies and other complex biological systems, local interactions between the parts are able to produce coordinated behavior of the whole without central control (see *Ant Encounters*, 19). Her research also shows how ants are not diligent, but "for the most part just hang around" (85), as interaction rates themselves act as important cues for the self-organization and maintenance of the colony.
    4. Latour, *Reassembling the Social*, 238.
    5. Bonabeau, Dorigo, and Guy, *Swarm Intelligence*, xi.
    6. Parikka, *Insect Media*, xxv.
    7. Parikka, xxv.
    8. Gordon, *Ant Encounters*, 1.
    9. Werber, "Schwärme," 198.
    10. Sleigh, *Six Legs Better*, 70.
    11. Sleigh, 163. Sleigh shows here in particular that "the experimental continuum posited by the cyberneticians from the biological to the social meant that ants were an obvious subject of scientific interest, thanks to their liminal status as organism and superorganism."
    12. Sleigh breaks the one hundred years of entomological research she examines (approximately from the 1860s to the 1970s) into three periods: "Ants metamorphosed through three main forms, appearing sequentially as psychological, sociological, and informational entities. In other words, they were used successively to model the human mind, society, and communication. For each period, one figure stands out from the scientific milieu. For the era of psycho-

logical modeling it is the Swiss psychiatrist Auguste Forel (1848–1931). For the sociological era it is the American academic and coiner of the term myrmecology, William Morton Wheeler (1865–1937), and for the information era it is the American sociobiologist Edward O. Wilson (1929–)" (*Six Legs Better,* 11). The last period directly draws on cybernetic and neocybernetic models of thinking.

13. Gordon, *Ant Encounters,* 3.

14. Sleigh, *Six Legs Better,* 25.

15. See Schrift, "Arachnophobe or Arachnophile?"

16. Already Aristotle noted that ant colonies have no leader, making them an ideal model for a self-governing republic. For an excellent overview of the history of ants and bees as models for political structures, see Johach, "Der Bienenstaat."

17. Thompson and Varela, "Radical Embodiment," 420.

18. Espinas, *Die thierischen Gesellschaften,* 351. Page numbers and markings in this chapter reference Nietzsche's personal copy of the book, which is available online via the Herzogin Anna Amalia Library (HAAB) of the Klassik Stiftung Weimar, haab-digital.klassik-stiftung.de/viewer/image/1236091299/1/LOG_0000/. I would like to thank the HAAB for having allowed me to review copies from Nietzsche's personal library that had not been scanned yet in 2017.

19. Nietzsche was likely familiar with the different readings of ants by Aristotle and Hobbes from Albert Lange's *Geschichte des Materialismus.* Lange describes how Hobbes replaces the "political instinct" that led Aristotle to equate humans and ants as "state-forming animals" with an absolutist theory of the state that escapes the *bellum omnium contra omnes* through a "superior will." "Not by virtue of political instinct," Lange summarizes Hobbes, "but through fear and reason man would acquire a unity among themselves for the purpose of shared security" (203).

20. In "Nietzsche and Networks, Nietzschean Networks," Dan and Nandita Biswas Mellamphy translate Nietzsche's distinction between Dionysus and Apollo into the terminology of information theory, offering a refreshing turn against the secondary literature's "strong tendency to resist interpreting Nietzsche outside the register of the 'organic' bias in his philosophy of life" (19). Unfortunately, they do so without contextualizing historically their own or Nietzsche's terminology, and without looking more comprehensively at the question of how Nietzsche's writings challenge anthropocentric and naturalizing modes of thinking.

21. Derrida, *Spurs,* 101. Derrida's spider analogy points toward the lack of control an author has over his writings, a fact Derrida sees acknowledged in Nietzsche's aphoristic, fragmentary, and poetic styles.

22. Lemm, *Nietzsche's Animal Philosophy*, 9.

23. Herbrechter, *Posthumanism*, 3.

24. In *The Wanderer and His Shadow*, Nietzsche first speculates that "the forest ant might imagine too that it is the aim and purpose of the forest's existence just as we do when we almost involuntarily connect the end of humanity with the end of the earth" (WS, 14). Only two aphorisms on, Nietzsche cites ants again, this time to relativize the human desire for metaphysical truths, suggesting that "humanity does not need certainty about the very last horizons to live a full and industrious life, just as the ant does not need it to be a good ant" (16).

25. If nothing else, contemporary post-truth culture makes evident how much the struggle over truth is a struggle about sociopolitical order. Chapter 7 will return to this question in the context of warnings about the dangers of Nietzsche's conception of truth by liberal thinkers.

26. Rhetorically, Nietzsche employs an apophasis here, a trope that allows him to "know" or say what he confesses cannot be known or said. The claim repeats *in nuce* the pragmatic stance of Nietzsche's epistemology that, as chapter 2 showed, has to acknowledge its own limits, including about what it "knows" are the limits of knowledge. In this regard, it is best to view Nietzsche's statements about nature, as Christian Bertino argues, as heuristic rather than ontological propositions ("Sprache und Instinkt," 81).

27. Leading up to the quote, Nietzsche stresses the idea of metaphoric freedom from conceptual slavery, suggesting: "So long as it is able to deceive without *injuring*, that master of deception, the intellect, is free; it is released from its former slavery and celebrates its Saturnalia. . . . With creative pleasure it throws metaphors into confusion and displaces the boundary stones of abstraction" (TL, 90).

28. Espinas, *Die thierischen Gesellschaften*, 351.

29. Maria Cristina Fornari examined the influence of Spencer's social theory on Nietzsche, who read Spencer in late 1879 and early 1880. Whereas psychological and moral social theories pursue "a free and responsible subject," Fornari shows that "evolutionist and utilitarian English and organicist sociologists, supporters of Spencerism (Alfred Fouillée, Jean-Marie Guyau, Alfred Espinas) seeking general well-being, impose alleged 'natural' purposes on the individual by integrating it in collective structures and turning it into a useful function of the whole, rather than aspiring to the real emancipation of the individual" ("Social Ties," 246). Fornari here points toward biology as the origins of the "idea of 'colonial' formations," as biology was "by now aware of the gregarious nature of tissues and cells." In an earlier essay tracing the significance of Spencer

for Nietzsche's moral philosophy, Fornari also argues that Fouillée and Espinas helped Nietzsche deduce morality from the predominance of drives ("Die Spur Spencers," 312).

30. Nietzsche's organicist sociology expands what Holub identifies as one of Wilhelm Roux's main arguments: that the "organism does not develop harmoniously, but by means of internal conflict that prepares it best for its environment" (*Nietzsche in the Nineteenth Century*, 341).

31. See Sleigh, *Six Legs Better*, 72. Sleigh also brings up the importance of Espinas's book for Émile Durkheim, noting parallels and differences: "Just as Espinas compared the anthill to the mammalian brain, so Durkheim analogized the society with divided labor to the body with its organs, specialized for their different tasks. The comparisons were functional and quasi-teleological. . . . If Durkheim saw society as a body, Espinas saw it as a mind. The true society, resulting from functional associations, was defined by Espinas as 'a living consciousness, or an organism of ideas.' He considered the ant colony to be 'truly, a single thought in action (albeit diffuse)'" (74–75).

32. While Nietzsche bought the German translation of Espinas's book in 1882, David Simonin finds evidence in Nietzsche's notes that indicates that Nietzsche knew about the main theses of the book already in 1880 ("Nietzsches Lektüre von Alfred Espinas," 301–2). Simonin's essay confirms what my analysis hopes to detail further and with specific reference to insect sociality, namely how Espinas helped Nietzsche break with the simple dichotomy between herd and individual, leading him instead to examine the "deep and problematic connections between both" (321).

33. See Nietzsche's markings at Espinas, *Die thierischen Gesellschaften*, 125.

34. See Nietzsche at Espinas, 128–30.

35. See Nietzsche at Espinas, 128.

36. See Nietzsche at Espinas, 128–29.

37. See Nietzsche at Espinas, 129.

38. For a popular, albeit ahistorical, account of phenomena of emergence across biological, social, and even technological planes, see Johnson, *Emergence*.

39. See Nietzsche's markings at Espinas, *Die thierischen Gesellschaften*, 128.

40. See Nietzsche at Espinas, 163.

41. See Nietzsche at Espinas, 166.

42. See Nietzsche at Espinas, 173.

43. See Nietzsche at Espinas, 175.

44. Holub's chapter "The Social Question" (*Nietzsche in the Nineteenth Century*, 125–72) documents Nietzsche's limited interest and limited familiarity

with socialist thinkers of his time, including with Marx, of whose writings Nietzsche most likely knew only indirectly from some lengthy footnotes in the second edition of Lange's *Geschichte des Materialismus*. On some of the likely sources for Nietzsche, see also Brobjer, "Nietzsche's Knowledge of Marx and Marxism."

45. As Holub points out in his comprehensive assessment of Nietzsche's stance on the "social question," the "problem for Nietzsche is not one of equality or breaking down hierarchies; it is one of sustaining hierarchical structures that are meaningful and that promote the nobility of spirit that can produce greatness" (*Nietzsche in the Nineteenth Century*, 158).

46. The next to last sentence of the aphorism speculates about the possible benefits Europe would derive from the supposed work ethics of the Chinese: "Indeed, they might as a whole contribute to the blood of restless and fretful Europe something of Asiatic calm and contemplativeness and—what is probably needed most—Asiatic *perseverance*" (D, 206). In a note from spring of 1888 entitled "Progress" that challenges the notion of social progress, Nietzsche again cites the Chinese as "a type that turned out well, that is, more durable than the European " (KSA, 13:408–9). For a comprehensive discussion of Nietzsche's image of China and some of his sources, see Cheung and Hsia, "Nietzsche's Reception of Chinese Culture."

47. Nietzsche disapproved of Bernhard Foerster's colonization plans as he disapproved of the nationalism and anti-Semitism that drove these plans. See, for example, Nietzsche's letter to Franziska and Elisabeth Nietzsche from March 14, 1885 (581, in KSB, 7:21–23).

48. See the entry for *schwärmen* in Jacob and Wilhelm Grimm, *Deutsches Wörterbuch*.

49. Deleuze and Guattari write that one "can never get rid of ants because they form an animal rhizome that can rebound time and again after most of it has been destroyed" (*A Thousand Plateaus*, 8). They also draw on insects, particularly the reciprocity in the relationship between wasps and orchids, to describe the kind of embeddedness implied by the concept of the rhizome (9).

4. Instinct, Will, and the Will to Power

1. Bennett, *Vibrant Matters*, x. Bennett's materialism thus wishes to acknowledge the "vibrancy of things" as "vivid entities not entirely reducible to the contexts in which (human) subjects set them" (5).

2. Hartmann, *Philosophie des Unbewussten*, 258. The first edition was published in 1869.

3. Quoted in Gerber, *Die Sprache als Kunst,* 190. Language, in Humboldt's assessment, emerges with and underlies an individuation process that apparently breaks the bonds of cooperating mental forces. ("Über die Verschiedenheit des menschlichen Sprachbaues und ihren Einfluss auf die geistige Entwicklung des Menschengeschlechts" was originally the introduction to *Über die Kawi-Sprache* [On the Kawi language], which was published posthumously, and this introduction was published by itself in 1836, and translated into English in the late twentieth century.)

4. Hartmann relates character to the ethical behavior of humans, which, he proclaims, "lies in the deepest night of the unconscious" (*Philosophie des Unbewussten,* 230).

5. Nietzsche slightly changes Hartmann's comparison of the formation of language with that of an anthill by stipulating that the instinct to build is "the most proprietary achievement of the individual *or* the mass" (my emphasis), while Hartmann drew on bees and the anthill to make the point that the origin of language cannot be explained in terms of the work of an individual, that "only the instinct of the mass can have created [language], just as [instinct] presides over the life of a beehive or a termite [hill] or anthill" (*Philosophie des Unbewussten,* 256). The change suggests that young Nietzsche assumes a reciprocity between the character of the individual and that of the mass.

6. Crawford, *Beginnings of Nietzsche's Theory,* 49.

7. Nietzsche's discussion of Kant's definition of instinct follows closely the reference to Hartmann in the notes to his unpublished "Lectures on Latin Grammar" (see KGW, 2/2:188).

8. See Bertino, "Sprache und Instinkt," 75.

9. Bertino, 75.

10. Parikka, *Insect Media,* 24.

11. Darwin, *Origin of Species,* 189.

12. Darwin, 191.

13. See Darwin, 192: "Changes of instinct may sometimes be facilitated by the same species having different instincts at different periods of life, or at different seasons of the year, or when placed under different circumstances, &c.; in which case either one or the other instinct might be preserved by natural selection." The malleability of instincts leads Alfred Espinas to reject the usefulness of resorting to the term altogether, arguing that "instinct" as an explanatory principle for animal behavior introduces a circular logic in which the instinct has to be put at the beginning of the process that produces it (*Die thierischen Gesellschaften,* 183).

14. See Klaus Rohde's entry "Instinkt" in Ritter et al., *Historisches Wörterbuch der Philosophie.*

15. See Holub, *Nietzsche in the Nineteenth Century*, for a comprehensive account of Nietzsche's philosophical engagement with these physiologists and dynamic monists. For a broader examination of the emergence of a concept of biological agency in nineteenth-century German cell theory and embryology, see Emden, "Agency without Humans."

16. See KSA, 10:314–15. The note is also in close proximity to the only mentioning of Espinas's name in Nietzsche's unpublished notes, which likens the police to birds protecting bovines from parasites (see KSA, 10:318).

17. Espinas, *Die thierischen Gesellschaften*, 351. Espinas is here discussing the organization of beehives, noting the absence of a "central power," marveling what "a strange state this is, in which there is not even the shadow of a government," and concluding that two basic drives, "motherly love and personal interest, suffice to create among thousands of individuals a harmonious cooperation without coercion." The passage is also quoted and discussed in Schneider, *Der thierische Wille*, 16–20.

18. In his *Vorlesungen über die Menschen- und Thierseele*, Wundt not only attributes to ants, termites and bees the ability to think and to act following considerations, but argues that they must have already means that allow them to communicate: "The appearance of social life certainly points toward the existence of a sign language: which can exist only if there is already a high degree of psychological development" (448). Wundt explicitly challenges the still-dominant opinion that instincts are purely mechanical reactions, a view he traces back to H. G. Reimarus's 1773 treatise *Allgemeine Betrachtungen über die Thiere, hauptsächlich über ihre Kunsttriebe* (490).

19. Cowan, "Nichts ist so sehr zeitgemäss als Willensschwäche," 50.

20. In a letter to Paul Rée from August of 1877, Nietzsche encourages his friend to collaborate with the English journal *Mind*, praising the quality of the quarterly, which collaborates with "all the greats of England," including Darwin, Spencer, and Taylor, and which has forthcoming a "great essay by Wundt on the 'Philosophy of Germany'" (Letter 643, in KSB, 5:265–67). Wundt's essay was being translated into English at the Swiss summer resort Rosenlaui, where Nietzsche was staying at the time and where he met the editor of *Mind*, who had noted Rée's writings. Incidentally, Wundt's essay mentions Nietzsche in not very flattering ways. Wundt suggests Nietzsche belongs to a group of philosophers who adhere to Schopenhauer, among whom "the pessimistic mood is combined in a very peculiar way with an enthusiastic devotion to certain ideas closely related to religious mysticism. Richard Wagner and his music are ardently worshipped by this sect of pessimists" ("Philosophy in Germany," 509).

21. On the surprisingly materialist background already of Schopenhauer's concept of the will, see Janaway's *Schopenhauer*.

22. Wundt distinguishes perception, the "entry of a representation into the inner visual field," from apperception which he defines as "its entry into the focal point [*Blickpunkt*]" (*Grundzüge*, 206).

23. Nietzsche was aware of Helmholtz's reflections on misrecognition (e.g., when we mistake clouds for a mountain range) as evidence of unconscious cognitive acts. Such common phenomena reveal the preconscious processing of our perceptions. They raise questions about the relationship between the physiological and mental determination of what we perceive. Sören Reuter sees Nietzsche engage Helmholtz's theory of unconscious judgments in the essay fragment *On Truth and Lies in a Nonmoral Sense*. Once we accept that perception does not reflect unchanging natural laws, but "habits that developed from life circumstances, . . . then sense physiology, viewed from this angle, analyzes basic types of human social behavior" ("Reiz—Bild—Unbewusste Anschauung," 370). Nietzsche, Reuter concludes, interpreted these habits of perception aesthetically.

24. See Wundt, *Grundzüge*, 85.

25. See Wundt, 218–19.

26. See Wundt, 387.

27. See Wundt, 391.

28. See Wundt, 393.

29. A similar dynamic is present in section 13 of the first essay of *On the Genealogy of Morals*, where Nietzsche derives slave morality from the tensions between oppressors and oppressed that led the latter to separate the doer from the deed. Chapter 6 will return to this passage.

30. Clark and Dudrick, "Nietzsche on the Will," 253. Cowan argues similarly that the "model of the good will always functions, in Nietzsche's later writings, as both an individual and a social model; . . . his philosophy of the will is always already a philosophy of the good social order. . . . And the central characteristic of such a well ordered and successful social formation, for Nietzsche, is precisely the clarity of roles between commanding and obeying instances, which is the single criteria for what he here calls, in the language of contemporary psychology, the 'state of pleasure' (*Lust-Zustand*) accompanying all acts of successful willing" ("Nichts ist so sehr zeitgemäss als Willensschwäche," 52–53).

31. Clark and Dudrick, "Nietzsche on the Will," 263.

32. See Clark and Dudrick, 252: "But in a case where we exercise willpower to overcome temptation, it seems plausible that the thought that directs our behavior takes the form of a command—e.g., 'Put down the fork this minute and

step away from that chocolate cake,' or, slightly nicer, when one really wants to keep reading The New York Times: 'Time to go back to work.'" It is not surprising that analytic philosophers would detect instances of circular logic in Nietzsche's conception of unconscious wills and drives, e.g., find "him explaining the activities characteristic of persons in terms of drives—which engage in the activities characteristic of persons" (265). It is surprising, though, that they ignore that Nietzsche was aware of such circularities and that they do not consider his response to them.

33. Abel, *Nietzsche*, 57. Abel offers a detailed analysis of the philosophical traditions, from Leibniz to Schelling to Schopenhauer, that shape Nietzsche's conception of the dynamics of competing wills-to-power from which the organic and inorganic world emerges (a verb Abel does not use) and which keep everything in constant flux.

34. Richardson, *Nietzsche's Values*, 186.

35. Richardson, 419. Chapter 7 will return to the anti-egalitarianism this stance implies and its extensive discussion in the recent Anglophone literature on Nietzsche's relevance for political theory.

36. Richardson, 232.

37. See Wundt, *Grundzüge*, 460–61. Wundt concludes that it is with consciousness that the world has come to contemplate itself ("*in welchem die Welt sich auf sich selbst besinnt*" [464]).

38. Kim, "Wilhelm Maximilian Wundt."

39. By "intelligible character," Nietzsche returns to the neurophysiological foundations of understanding we saw in the ant-citing note from 1883. In aphorism 36 of BGE, he relates thinking to the drives and passions: "Assuming that our world of desires and passions is the only thing 'given' as real, that we cannot get down or up to any 'reality' except the reality of our drives (since thinking is only a relation between these drives)." See also KSA, 5:54–55.

40. Pippin, *Nietzsche, Psychology, and First Philosophy*, 4–5. For a critique of the "historical misunderstanding" of Nietzsche's psychology, which "refers not to cognitive psychology but to the psychological sciences of the nineteenth century," see Emden, "Nietzsche's Will to Power," 32. Emden suggests that Nietzsche's concept of a will to power as developed in aphorism 36 of BGE, "despite its seemingly metaphysical ramifications, tries to achieve precisely that: naturalizing the preconditions of naturalism" (37). Ian Dunkle argues similarly that "Nietzsche offers the [will to power] as a theory of life primarily for the purpose of defending his moral psychology against competing accounts that claimed a purely biological basis" ("Moral Physiology," 80).

41. Bennett, *Vibrant Matters*, x. While Bennett's materialism builds on the "material vitalism" Deleuze and Guattari developed in their "1227: Treatise on Nomadology—A War Machine" in *A Thousand Plateaus*, she also situates her materialism within a larger tradition of vitalist thought that reaches from Epicurus, Lucretius, Thomas Hobbes, Baruch Spinoza, and Denis Diderot to Nietzsche, Henry David Thoreau, Henri Bergson, and Hans Driesch.

42. Cowan, *Cult of the Will*, 6.

43. Cowan, 14.

44. Cowan, 14.

45. Nietzsche uses the term "disgregation"—a term he borrows from physics, where it designates the thermal separation of the molecules of a body—in his published works once in *The Case of Wagner* and twice in *Twilight of the Idols* (1889). It first appears in his notes in the Fall of 1885 (see KSA, 11:701–2). Hamacher focuses mainly on Nietzsche's use of "disgregation of will" in his discussion of *décadence* in the *Case of Wagner* 7. In *TI*, Skirmishes 35, Nietzsche speaks of the "disgregation of instincts," which he associates with altruism, and by extension with nihilism.

46. Hamacher, "Disgregation des Willens," 309.

47. Hamacher, 310.

48. Abel, *Nietzsche*, 134–35.

49. Hamacher, "Disgregation des Willens," 324. In *Orientierung im Nihilismus*, Werner Stegmaier argues similarly, quoting aphorism 19 of BGE, that, if "willing is to be understood as a wanting-to-subject-an-other, on the one hand, and as wanting-to-place-yourself-in-an-other, on the other, [then] 'a philosopher should claim the right to understand wanting itself within the framework of morality: morality understood as a doctrine of the power relations under which the phenomenon of "life" arises'" (247). In GS 354, Nietzsche makes a similar argument. Linking the development of language to the development of consciousness, he notes that "consciousness actually belongs not to the individual-existence of a human, but rather to the community- and herd-nature in him." John Richardson notes the surprising consequence of Nietzsche's argument: "Consciousness, language, and . . . agency originated to serve society's interests, not those of the (original) drives and affects" (*Nietzsche's Values*, 191).

50. Hamacher, "Disgregation des Willens," 322.

51. Hamacher, 324.

52. GS 149 is entitled "The Failure of Reformations" and suggests that the failing of the Greeks' attempts to found new religions "indicates that even early in Greece, there must have been many diverse individuals whose diverse plights

could not be disposed of with a single prescription of faith and hope." Abel, too, notes that, for Nietzsche, disgregation and cultural decline, viewed from the inside, must also be interpreted as a form of gain (*Nietzsche,* 118).

53. Lemm, *Nietzsche's Animal Philosophy,* 154. Lemm suggests that Nietzsche "proposes to investigate culture as part of the continuum of life, as constituted out of animal life" (155). Which is precisely what Nietzsche does when he derives language and understanding, as noted earlier, from rudimentary neurophysiological responses to social interactions that cut across species.

54. Hamacher writes: "The connection of those who have lost or are about to lose all connection, cannot be imagined other than as the product of a fiction of connectivity: for example under the fiction of the concept of the individual" ("Disgregation des Willens," 325).

55. Derrida continues: "Hence this logic of force bows to a law stronger than that of force. The logic of force reveals within its logic a law that is stronger than this very logic. . . . One must defend the weakest who are pregnant with the future, because it is they who are the strongest" (Derrida and Beardsworth, "Nietzsche and the Machine," 31).

56. For reflections on the figure of the Archimedean point in modern philosophy and contemporary theory, see Holland and Landgraf, *The Archimedean Point in Modernity,* and therein with regard to Nietzsche, see Landgraf, "Circling the Archimedean Viewpoint," 88–106.

5. Media Technologies of Hominization

1. Wolfe, *What Is Posthumanism?,* xiii.

2. While Nietzsche is identified like no other philosopher with posthumanism in film and on TV, as Babette Babich points out, unfortunately "the popular mind's tendency [is] to favour a one-dimensional characterization of Nietzsche [which] affects even scholars . . . who write on Nietzsche and transhumanism" ("Friedrich Nietzsche and the Posthuman/Transhuman," 28).

3. Leo Marx offers a concise overview of technology's semantic history in the nineteenth century and how technology is recognized as a historical force on its own in the early twentieth century, as "the idea of progress had become the fulcrum of a comprehensive worldview effecting the sacralization of science and the mechanical arts, and creating a modern equivalent of the creation myths of premodern culture" ("Technology," 565). He appears to reaffirm the assumption of a fundamental heterogeneity between humans and technology, however, when he identifies the "hazardousness" of the concept with how it "distracts at-

tention from the human—socioeconomic and political—relations which largely determine who uses them and for what purposes" (576).

4. An important exception to the popular rendering of the cyborg is offered by Donna Haraway's famous "Cyborg Manifesto," which approaches the cyborg from a poststructuralist angle, exploring the conceptual impact of the figure on gender and other norms. I will return to Haraway's approach at the end of this chapter.

5. Vanessa Lemm argues that a philosophical anthropology that understands humans along Helmuth Plessner's line as deficient beings (*Mängelwesen*) is incompatible with Nietzsche's anthropology because "insecurity of the human with respect to other forms of life is no longer the fundamental point of Nietzsche's philosophical anthropology" (*Homo Natura*, 172). I will return to Lemm's observation below.

6. We will return to the only prominent engagement with technology in Nietzsche below, which is Friedrich Kittler's media-technological reflections on Nietzsche's use of the typewriter. Much of the remaining literature on the topic adopts a pessimistic view that focuses on instances where Nietzsche seems to suggest that technology compromises human nature. Geoff Waite's book *Nietzsche's Corps/E*, despite its subtitle and an expansive chapter on the "Spectacular Technoculture of Everyday Life" (339–90), fails to offer a careful analysis of Nietzsche's possible contributions to a philosophy of technology. The chapter focuses instead on more recent appropriations of Nietzsche by neo-Marxist and other left-leaning thinkers. Summarizing Nietzsche's general stance on technology, but without reflecting on the term itself, Tarmo Kunnas claims that Nietzsche had a split view on it: as revealing "in a very concrete way the alienation of modern European culture," and yet (Kunnas speculates) holding hope for "a complete change of the whole of culture and civilization. It even might lead to a new cultural heyday in the spirit of Nietzsche's overman" ("Nietzsche und die Technologie," 317, 319). Arthur Kroker does not challenge the nature–technology divide when he outlines the ill effects of technology described by Heidegger, Nietzsche, and Marx, only to conclude that we have not grasped deeply enough yet "how graphically, how bleakly we truly have become a culture of the post" (*The Will to Technology*, 5).

7. Stiegler, *Technics and Time*, 141.

8. Stiegler, 141. As the field of paleoanthropology continues to evolve, so does today's understanding on how tools evolved along with humans and vice versa. For two more recent examples reflecting on the evolution of the human brain and the human hand respectively, see Stout et al., "Neural Correlates of

Early Stone Age Toolmaking," and Williams-Hatala et al., "Manual Pressures of Stone Tool Behaviors." For connections between the development of the human physiology (including the brain) and cooking, see also Wrangham, *Catching Fire*.

9. Stiegler, *Technics and Time*, 137.

10. Winthrop-Young, "Cultural Studies," 93–94.

11. Wolfe has expanded on the shortcomings of cultural studies from the point of view of an animal studies that would do more than merely extend the pluralism promoted by cultural studies "to previously marginalized groups without in the least destabilizing or throwing into question the schema of the human who undertakes such pluralization," one that would "bring [cultural studies] to an end . . . because it fundamentally challenges the schema of the knowing subject and its anthropocentric underpinnings sustained and reproduced in the current disciplinary protocols of cultural studies" ("Human, All Too Human," 568). In analogy, this chapter adopts a notion of technology that challenges the naturalization of "the human" as it challenges the very separation of technology from humans and their evolution.

12. Siegert and Winthrop-Young note connections and differences between this new generation of German scholars of *Kulturtechniken* and their American counterparts, the critical posthumanists Wolfe, Haraway, David Wills, and N. Katherine Hayles. Siegert posits the main difference as the increased focus on media—indicative of the influence of Kittler—and a lack of political ambition on the German side: "While the American side pursues a deconstruction of the anthropological difference with a strong ethical focus, the Germans are more concerned with technological or medial fabrications or artifices. From the point of view of the cultural techniques approach, anthropological differences are less the effect of a stubborn anthropo-phallo-carno-centric metaphysics than the result of culture-technical and media-technological practices" ("Cultural Techniques," 55).

13. Fuchs, "Circularity of the Embodied Mind," 7.

14. Lemm is quoting Braidotti, "Posthuman Critical Theory," 19.

15. As an example for the slippage between the psychological, social, and physiological effects of technology, consider recent research into the impact of digital game-play that led the World Health Organization to add "Gaming Disorder" to the International Classification of Diseases (ICD) ("Addictive Behaviours: Gaming Disorder").

16. See Dünkelberg, *Encyclopädie und Methodologie der Kulturtechnik*.

17. For a concise elaboration of the distinction between *epistêmê* and *technê* throughout ancient Greek philosophy, see Richard Parry's entry "*Episteme* and

Techne" in *The Stanford Encyclopedia of Philosophy*. Parry makes the point that some of the contemporary lines we draw between theory and practice that make the opposition seem irreconcilable primarily go back to Aristotle, but do not necessarily apply to the relation between *epistēmē* and *technē* in other Greek philosophers, including Plato.

18. See Lambrecht, *Lehrbuch der Technologie*. For a concise summary of Lambrecht's "textbook" and the conceptual challenges it faced, see Holland, "Instruction in an Imperfect Science."

19. See Beckmann, *Anleitung zur Technologie* and *Entwurf einer allgemeinen Technologie*.

20. Marx, *Capital: Volume 1*, 375. In the German of the first volume *Das Kapital*, the note reads: "*Die Technologie enthüllt das aktive Verhalten des Menschen zur Natur, den unmittelbaren Produktionsprozess seines Lebens, damit auch seiner gesellschaftlichen Lebensverhältnisse und der ihnen entquellenden geistigen Vorstellungen*" (357).

21. Spivak, *Critique of Postcolonial Reason*, 77. In his 1963 essay on "Marxism and Humanism," Louis Althusser had established already that, in 1845, "Marx broke radically with every theory that based history and politics on an essence of man" (section III).

22. Leif Weatherby recently profiled the tradition of a Romantic organology in Marx's work, where "the human becomes the organ of the machine; the machine is the collapse of both organics and mechanics and physical and intellectual labor" (*Transplanting the Metaphysical Organ*, 347). Weatherby reads Marx's *Capital* as an attempt to respond to the new kind of thinking that big industry introduced in the nineteenth century. It is not just the working conditions in nineteenth-century factories that require a new (material) philosophy, but the conceptual confusions and blind spots they created that Marx needed to address: "The point, however, is that the *Ungeheuer* [monster] is not merely the machine, like the famous monster that swallows lines of workers in Fritz Lang's *Metropolis*. It is the specific combination of relations and means, of form and instrument, that has spun out of control. Any thought based on notions of individual-social interface will fail to grasp this problem in its material urgency. When machines externalize systems of production, schematizing cherished metaphysical patterns and engraving those patterns on the social, new thought is called for. For that, the organological strain of *Capital* was needed" (343).

23. Marx, *Capital: Volume 1*, 375.

24. Marx, *Capital: Volume 1*, 378.

25. Schatzberg, "*Technik* Comes to America," 509. Schatzberg finds this con-

ception of technology fully spelled out first in Charles A. Beard's 1927 "Time, Technology, and the Creative Spirit in Political Science." Schatzberg suggests that what makes Beard's understanding of technology innovative is in part that "Beard divorced technology from capitalism, insisting that its influence was independent of any specific economic system" (510). While Marx might not have been consistent on this point, the quotes above suggest that Marx had already considered the possibility of technology functioning as a central agent of historical change.

26. Hörl, "Introduction to General Ecology," 13.
27. Marx, *Grundrisse*, 699.
28. Marx, 706.
29. Marx, "Economic and Philosophical Manuscripts (1844)," 362.
30. Marx and Engels, *Manifesto of the Communist Party*, 16.
31. Kapp, *Elements of a Philosophy of Technology*, 43.
32. See Kapp, 50: When a human being concerns "itself with the stone and, through the repeated grasping and inspection of it, begins to adapt it for use by his hand, then he is providing himself with it, arming himself with it. . . . The stone-throwing baboon repeats the same operation that it has for millennia, while the stone thrown by the primitive human was already the promise of the tool and an entire world of machines."
33. Plessner, *Gesammelte Schriften*, 4:382.
34. Stiegler, *Technics and Time*, 137.
35. In *Thus Spoke Zarathustra*, Nietzsche uses the caricature of the "ultimate" or "last human" (*letzter Mensch*) to draw a stark contrast to the overhuman. The former embodies what Nietzsche suggests is the ultimate ideal of the values produced by the humanist tradition, a life that seeks first and foremost comfort and repose over challenge and self-overcoming. This "last human" is only one of a series of humans that include the "free spirit" and the "sovereign individual," who, Nietzsche argues, evolved from Western society's cultivation processes, processes that he sees coming to a head in his time.
36. Cited in Kittler, *Discourse Networks*, 196.
37. Kittler, *Discourse Networks*, 178.
38. Wellbery, "Foreword," xxxi.
39. Kittler, *Discourse Networks*, 194.
40. Kittler, 196.
41. Moore, "Introduction," 2. See also Laura Otis, *Networking*, on the development of these technologies and their impact on language—and hence philosophy—at the time.

42. Language, as noted in the previous chapter, is not what separates humans from the animal, as Nietzsche conceives of language as a product of "nature," of instinct, not as its opposite. Christian Bertino specifies that Nietzsche links human self-consciousness to the communicative use of language, rather than vice versa, as Herder does ("Sprache und Instinkt," 70).

43. The invention of the rotary press in 1843 (it reached Germany in 1873, the year Nietzsche published his *Untimely Meditations*) accelerated the printing of newspapers in the second half of the nineteenth century. Nietzsche continued to underline the newspapers' ill effects on culture, including in the preface to BGE and in aphorism 263 therein.

44. Large, "Nietzsche, Burckhardt, and the Concept of Culture," 10.

45. Quoting himself as a "forgotten" authority, in the second *Untimely Meditation* Nietzsche repeats his definition of culture from the first.

46. As noted in chapter 4, Nietzsche understands human nature expressed in instincts, passions, habits, as preconscious and as continually changing. Immediately preceding the quote, he writes: "For since we are the outcome of earlier generations, we are also the outcome of their aberrations, passions and errors, and indeed of their crimes; it is not possible wholly to free oneself from this chain. If we condemn these aberrations and regard ourselves as free of them, this does not alter the fact that we originate in them. The best we can do is to confront our inherited and hereditary nature with our knowledge, and through a new, stern discipline combat our inborn heritage and implant in ourselves a new habit, a new instinct, a second nature, so that our first nature withers away" (HL, 76).

47. Kittler focuses on these and other passages in Nietzsche's work that lament how reading has become a lost art, how modern readers "improvise some approximation" (BGE, 192) of the text, rather than ruminate as he recommends for the digestion of his own writings. Nietzsche also blames the critical thrust of the Enlightenment when he observes that the "work never produces an effect but only another 'critique'; and the critique itself produces no effect either, but again only a further critique" (HL, 87).

48. See Baecker, *Wozu Kultur?*, 9–10. Baecker adds that the concept of culture was more successful than any other in positing the contingency of all formerly unquestioned pursuits of society while at the same time offering values intended to hide this achievement. For a more in-depth discussion of what Nietzsche identifies as the disintegration of modern culture, see Landgraf, "Disintegration of Modern Culture."

49. Heide Schlüpmann understands Nietzsche as distinguishing between cultures and "cults of knowledge" whereby the word "cult" is read as entailing the

appropriation of past cultures. See Schlüpmann (*Friedrich Nietzsches ästhetische Opposition*, 43). I would argue instead that, because premodern cultures did not recognize different cultures as cultures, they acted like cults. Western modernity, however, lost the belief in its cultural practices, and thus came to recognize their contingency.

50. In Germany, these hierarchies had been politically and nationally coded since the eighteenth century as the lower values were associated with French courtly culture, which was deemed overly civilized rather than cultured. See Elias, *Civilizing Process*, 5–44.

51. Bateson, "Form, Substance, and Difference," 453.

52. Herbrechter, *Posthumanism*, 21.

53. Graham, "Nietzsche Gets a Modem," 65.

54. Graham, 66.

55. Graham, 66, 72. Graham specifies further that "transhumanism owes a great deal to eighteenth and nineteenth century humanism, not only in its technophilic embrace of the prospects of scientific innovation, but in its vision of humanity freed of the constraints of superstition, ignorance and fear and liberated to pursue a brilliant destiny" (69).

56. In *The Singularity Is Near*, Ray Kurzweil frequently cites Nietzsche's concepts of the overhuman and of "overcoming," without however offering anything but a surface reading of his philosophy. Transhumanists appear largely to agree that Nietzsche "seemed not to see a role for technology" (More, "The Overhuman in the Transhuman," 3) in the transformation of humans, a view that owes itself, as I argued above, as much to a rather narrow definition of technology as to a highly selective reading of Nietzsche's *Zarathustra* that fails to account for the attention Nietzsche pays to the role material, linguistic, and media communication technologies play in hominization and cultivation processes.

57. Haraway, "Cyborg Manifesto," 118, 120.

6. Cultivating the Sovereign Individual

1. Sharon, *Human Nature*, 5.

2. Braidotti, *The Posthuman*, 102, 136.

3. Braidotti, *The Posthuman*, 136.

4. See Emden's seminal book *Nietzsche's Naturalism*, as well as his article "Nietzsche's Will to Power." Helmut Heit offers an overview of the history of naturalization attempts of Nietzsche's philosophy and engages the contemporary debates around this topic, arguing that "there is no consensus regarding scope,

version and justification of his apparent naturalism" and that "much dispute therefore centers around the appropriate qualification of Nietzsche's naturalism such as 'ontological,' 'methodological,' 'speculative,' 'aesthetic,' 'perspectival,' 'perfectionist,' 'anti-scientistic,' 'normative,' etc." ("Naturalizing Perspectives," 63). At the end of this chapter, I will return in more detail to Vanessa Lemm's book *Homo Natura*, which raises the question of Nietzsche's naturalism specifically in reference to a discussion of his posthumanism.

5. Emden, *Nietzsche's Naturalism*, 6.
6. Braidotti, "Posthuman Critical Theory," 29.
7. See chapter 15 of Ferrando, *Philosophical Posthumanism*, entitled "Technologies of the Self as Posthumanist (Re)Sources" (82–84).
8. Braidotti, "Theoretical Framework," 31.
9. Barad, *Meeting the Universe Halfway*, 352.
10. Haraway, *Staying with the Trouble*, 58.
11. For a multifaceted discussion of the role agonistic relations play in Nietzsche, see Siemens and Pearson, *Conflict and Contest*. Agonistic relations do not preclude the possibility of "respectfulness," in fact they require it, as William Connolly suggests with his concept of "agonistic respect." In the concluding chapter of this book, I will return to the debate about the political implications of this aspect of Nietzsche's thought and the question of how his posthumanism might support or even be able to revitalize democracy, as scholars such as Siemens, Acampora, Hatab, and Shrift argue.
12. As John Richardson notes specifically with reference to Nietzsche's perspectivism: "Perspectives do not stand all on the same level but rather higher and lower than one another" (*Nietzsche's Values*, 461). Richardson suggests that Nietzsche's elitism is nevertheless meant to improve the lot of those of lower rank: "Nietzsche thinks that the elite's 'paternalism' will manifest in an effort to give the herd norms that aim members better at their own growth" (473).
13. Seeking power, of course, should not be equated with seeking violence, but as Paul Katsafanas puts it, "with the activity of perpetually seeking and overcoming resistance to one's ends" ("Philosophical Psychology," 300). Katsafanas reads Nietzsche's concept of the will to power as "uncovering the deep structure of drives"—why the will to power for Nietzsche can "serve as a principle of revaluation of all values" (311).
14. Patton, "Nietzsche on Rights," 486.
15. Dries, "Early Nietzsche," 49. For an excellent summary of how assessments of Nietzsche's understanding of body, soul, and subjectivity have changed in recent years, see Guerreschi, "Leib, Seele und Subjektivität." The argument in

this chapter will be narrower and focus on how Nietzsche reconstructs the humanist creation of the self in *On the Genealogy of Morals* and in what respect this narrative *ex negativo* points toward a posthuman self.

16. Butler, "Foucault and the Paradox of Bodily Inscription," 602.

17. Kittler, *Discourse Networks*, 196.

18. Hayles, "Cognitive Nonconscious," 181.

19. Though it is not essential to our focus on the technicity of language, I agree with Andrew Huddleston that Nietzsche's genealogical account describes not, as most readers interpret the revolt, "a cunning ploy on the part of the weak and oppressed 'slaves' themselves to topple 'the nobles' from power—with the invention of morality being the slaves' weapon of choice," but that the slaves "are in fact the pawns of those Nietzsche calls 'the priestly people,' who foment this revolt by creating a new religion—Christianity—and a new system of values—morality—and foisting them on nobles and slaves alike, who in turn eventually come to organize their lives by these new ideals" ("Consecration to Culture," 137).

20. Pippin, "Lightning and Flash, Agent and Deed," 139.

21. Pippin, 140.

22. Owen, "Nietzsche, Ethical Agency and the Problem of Democracy," 145.

23. Owen, 148–54.

24. Kant, *Critique of Judgment*, 174 (§46).

25. See Luhmann, *Art as a Social System*, 203–4: "In this sense, creating a work of art—according to one's capabilities and one's imagination—generates the freedom to make decisions on the basis of which one can continue one's work. The freedoms and necessities one encounters are entirely the products of art itself; they are consequences of decisions made with the work. The 'necessity' of certain consequences one experiences in one's work or in the encounter with an artwork is not imposed by law but results from the fact that one began, and how." For a more detailed reflection on the problem of agency in the aesthetics of autonomy and in improvisation, see Landgraf, *Improvisation as Art*, and with specific reference to the question of agency in posthumanism and critical improvisation studies, Landgraf, "Improvisation, Posthumanism, and Agency in Art."

26. Kapp, *Elements of a Philosophy of Technology*, 202. Kapp subsequently puts language in the middle of a dialectical process between head and hand that produces self-consciousness: "Self-consciousness could not have developed if head and hand had not continually moved to accommodate one another—that is to say, if technology had not supplied stimuli and substance to the need for

language and if language had not imparted to the need to give form conscious awareness of its capacities and how to perfect them" (208).

27. Kapp, 203.

28. Wolfe, *What Is Posthumanism?*, xxvi.

29. Wolfe, xxvi. If we were to follow Sharon's cartography, Luhmann does not fall into the radical, politically motivated side of the posthumanist divide, but aligns with Bruno Latour and similar approaches that emphasize how human agency is socially and materially conditioned. Nevertheless, the proximity Wolfe notes between Derrida and Luhmann points toward a level of reflexivity on the non-essentializing side of posthumanism that is unique not only in how it "does not turn away from the complexities and paradoxes of self-referential autopoiesis" (xxi), but also in the ethos it carries. I will expand on the latter in the last chapter.

30. Wolfe, "Human, All Too Human," 571. In the essay, Wolfe discusses more extensively than in his book on posthumanism how the technicity of language in Derrida challenges the self-presence and authority of "the human."

31. Richardson, *Nietzsche's Values*, 168.

32. Richardson, 172.

33. Siemens, "Nietzsche's Socio-Physiology of the Self," 639. Siemens links the nonsovereignty of the self to the dominance he sees Nietzsche attribute to the state and to society. Though Siemens treats the latter as independent entities rather than as evolutionary aggregates, he nevertheless reaches the conclusion that "Nietzsche's socio-physiology forbids the abstraction of our capacity to reason from our affective, embodied existence. Not only are our 'experiences and judgements' incorporated and learnt from the state; so too are our very affects and drives. Together, they are pre-formed by the interests and functions of the social organism to which we originally belong. This rules out not only those liberal contract theories that presuppose our capacity for reason or autonomous reflection (e.g. Rawls), but also those that presuppose primordial affects and drives on the individual's part, such as Hobbes' fear of death and desire for self-preservation" (637).

34. Meredith, "Bound Sovereignty," 236. Meredith offers a concise summary of the debate in the secondary literature on the purported sovereignty of the sovereign individual, concluding that "both Nietzsche's analysis of the sovereign individual and his genealogy of conscience suggest that autonomy is unlikely if not impossible within the political community" (241).

35. That this "ripest fruit" is not an ideal we should emulate is also backed up by Nietzsche's prior use of the metaphor in the second *Untimely Meditation*,

where he clearly distances himself from the figure, chastising the false pride of those who think, "we have reached the goal, we are the goal, we are nature perfected" (HL, 108). In "Breaking the Contract Theory," Hatab also argues that the sovereign individual must itself be overcome.

36. Latour writes that "we might call technology the moment when social assemblages gain stability by aligning actors and observers. Society and technology are not two ontologically distinct entities but more like phases of the same essential action" ("Technology Is Society Made Durable," 129). Latour's notion of technology aligns with how Nietzsche extends interpretive schemas and the institutions they subtend to the social realm.

37. Hatab, "Breaking the Contract Theory," 181.

38. When Zarathustra, in the same section, identifies as tarantulas and poisonous spiders the "preachers of equality," he is, as Babette Babich argues, "not arguing against social equality because he is against the idea but because the claim is impossible, corresponding to a shortsightedness that happens to be ideologically self-serving" ("Towards Nietzsche's 'Critical' Theory," 123).

39. See Lende and Downey, *Encultured Brain.*

40. Wolfe, "Human, All Too Human," 568. I am extending here what Wolfe, citing Tilottama Rajan, claims for certain strands of popular animal studies, to include the role technology continues to play in certain strands of posthumanism and transhumanism. Wolfe himself points in the same article toward the intersection of animal studies with posthumanism, "not in the sense of some fantasy of transcending human embodiment, but rather in the sense of returning us precisely to the thickness and finitude of human embodiment and to human evolution as itself a specific form of animality, one that is unique and different from other forms but no more different than an orangutan is from a starfish" (572).

41. Hatab, "Talking Ourselves into Selfhood," 183.

42. Lemm, *Homo Natura,* 171, citing Braidotti, "Posthuman Critical Theory," 19.

43. Braidotti, "Posthuman Critical Theory," 24.

44. Braidotti, "Theoretical Framework," 33.

45. Braidotti, "Posthuman Critical Theory," 19.

46. See also Braidotti, 17: "What is needed instead is careful negotiation in order to constitute new assemblages or transversal alliances between human and nonhuman agents, while accounting for the ubiquity of technological mediation. My argument is that we need to take the challenge of transformation right into the fundamental structures of subjectivity: the posthuman turn is not to be taken for granted."

47. Lemm, *Homo Natura*, 172.
48. Lemm, 102.
49. Butler, "Foucault and the Paradox of Bodily Inscription," 602.
50. Lemm, *Homo Natura*, 172.

7. The Ethics and Politics of Nietzschean Posthumanism

1. Siemens and Roodt, *Nietzsche, Power and Politics*, 1.
2. Siemens and Pearson, *Conflict and Contest*, 57.
3. Siemens and Pearson, 57.
4. Sharpe, "Golden calf," 71. For warnings about the danger Nietzsche presents to liberalism, see also Malcom Bull's *Anti-Nietzsche* and Ronald Beiner's *Dangerous Minds*.
5. "Nietzsche's hatred of 'Jew hatred'" is the title of Brian Leitner's review of Robert C. Holub's 2015 *Nietzsche's Jewish Problem*. Holub reopened the debate, detecting a negative bias in Nietzsche's stance toward Jews and Judaism despite his many rejections (published and in letters) of anti-Semitism and anti-Semites. Leitner counters that arguing "against moralities endorsed by Christians, Muslims, and Jews is not a case of wrongful anti-Jewish prejudice" and that if "one *really* reads Nietzsche in context, what is striking is that the genuinely anti-semitic vitriol with which he was surrounded (and which Holub powerfully documents) made no systematic impact on his work and, indeed, came in for much mockery."
6. Warning about the dangers of questioning the contours of the human, Lyotard points to Nietzsche being "taken hostage by fascist mythology" (*The Inhuman*, 1). For an extensive account of connections between Nietzsche and National Socialism, see Galindo, *Triumph des Willens zur Macht*.
7. Beiner, *Dangerous Minds*, 12.
8. Jürgen Habermas describes Nietzsche's philosophy as a "turning platform" (*Drehscheibe*), implying that it derailed the Enlightenment train's progress. The railroad metaphor is part of the German title of the fourth lecture of Habermas's book *The Philosophical Discourse of Modernity*, entitled "Eintritt in die Postmoderne: Nietzsche als Drehscheibe" (Entry into postmodernity: Nietzsche as turning platform).
9. Patton, "Recent Work," 383. Patton also mentions Frederick Appel, Bruce Detwiler, Don Dombowsky, and Mark Warren as contemporary commentators who read Nietzsche as a proto-Fascist (385).
10. Ferrando, *Philosophical Posthumanism*, 151.
11. Braidotti, "Posthuman Critical Theory," 27.

12. Beiner, *Dangerous Minds*, 17.

13. Beiner, 7.

14. Latour, "Technology Is Society Made Durable," 128.

15. Beiner laments that "generations of readers of Nietzsche have never failed to find ways to 'launder' or 'sanitize' or at least take the edge off his hatred for freedom and equality" (18) and finds it to be a "curious and somewhat bizarre fact that Nietzsche is widely celebrated as the very archetype of an 'anti-foundationalist' and 'post-metaphysical' style of philosophizing" (40). Not surprisingly—but problematically, considering that the far-right appropriators who are the target of his critique, do the same thing—Beiner neither engages the existing Nietzsche scholarship in any comprehensive manner nor offers any specifics about "anti-foundationalism" and its political implications.

16. Patton, "Recent Work," 385.

17. Drochon, "Old Carriage," 1056.

18. Chantal Mouffe's *On the Political* adapts a point most prominently raised by Carl Schmitt, who warned against the danger of universal humanistic ideals being politically (ab)used to justify inhumane actions. On the democratic virtues of antagonistic rather than consensual political models, see also Rebentisch, *Art of Freedom,* and Rasch, *Sovereignty and Its Discontents.*

19. Drochon, "Old Carriage," 1056.

20. Miyasaki, "Nietzschean Case for Illiberal Egalitarianism," 155.

21. Holub, *Nietzsche in the Nineteenth Century,* 172. Holub notes that Nietzsche's "position entails an opposition to the evolving capitalist economy as such and encompasses a disdain for both the oppressor and the oppressed" (150).

22. Müller, *What is Populism?,* 20.

23. The slippage in Nietzsche's writings—that in the late 1880s, he increasingly adopts and exaggerates the dichotomies he deconstructs—has often been noted. It is evident in particular with regard to his take on democracy, which in *Human, All Too Human,* for example, is still quite differentiated and circumspect, but becomes increasingly agitated and provocative in *Beyond Good and Evil* and *On the Genealogy of Morals.* It also applies to Nietzsche's idea of self-overcoming, which with the concept of the *Übermensch* aims at a liberation from metaphysical and moral traditions of thought potentially open to anyone; but in his later writings they take a more exclusive and elitist form that led Nietzsche to make statements that, as Stegmaier puts it, "after the humanitarian catastrophes of the twentieth century have become completely inacceptable" (*Orientierung im Nihilismus,* 167–68).

24. Wolfe, "Ecologizing Biopolitics," 218.

25. Wolfe, 218.
26. Butler, "Foucault and the Paradox of Bodily Inscription," 602.
27. Hörl, "Introduction to General Ecology," 32.
28. Miyasaki who states explicitly that he is proposing "a Nietzschean defense of egalitarianism, not Nietzsche's" ("Nietzschean Case for Illiberal Egalitarianism," 155).
29. See Müller, *What Is Populism?*, 19–20.
30. Ure, "Resentment/Ressentiment," 600. Considering the potential utility of resentment, Ure suggests we distinguish between useful sociopolitical resentment and dangerous ontological resentment and reads Nietzsche as challenging "democratic political theory to consider whether or how we might prevent sociopolitical resentment from degenerating into ontological ressentiment" (607), in order "to avoid totalitarian or perfectionist politics" (610). While the distinction is helpful to account for the relevance of emotions in existing democracies, it is hard to separate the term "resentment" from the psycho-economic calculus that I argue a critical posthumanist ethos would want to overcome.
31. Luhmann, "Technology, Environment and Social Risk," 299.
32. Luhmann, 230.
33. Braidotti, "Posthuman Critical Theory," 27.
34. Braidotti's ethics are more Nietzschean when she builds on "Deleuze and Guattari's conception of anti-humanism as a function of 'becoming animal,'" which Lemm suggests "is clearly one of the most heavily Nietzsche-dependent conceptual constructions of the French philosophers" (*Homo Natura*, 173). In "Posthuman Critical Theory," Braidotti identifies as the "key notion in posthuman nomadic ethics" the "transcendence of negativity" that leads her to argue that "the conditions for renewed political and ethical agency . . . have to be generated affirmatively and creatively by efforts geared to creating possible futures, by mobilizing resources and visions that have been left untapped and by actualizing them in daily practices of interconnection with others" (27).
35. Lemm, *Homo Natura*, 176, quoting Wolfe, *Before the Law*, 55, and Esposito, *Bios*, 89.
36. Lemm, *Homo Natura*, 176–77.
37. Lemm, 180. Lemm had already argued in 2014 that Nietzsche's idea of resistance "presuppose[s] a prior relation to an other" and that a consideration of the other would "lead us beyond an individualist account of agency and power toward a relational account of (will to) power in Nietzsche" ("Shedding New Light," 351).
38. Braidotti, "Posthuman Critical Theory," 29. Braidotti writes further here:

"'We' become posthuman in this awareness of what no longer is the case: a unitary definition of the human sanctioned by tradition and customs. But we do remain human and all-too-human in the realization that the awareness of this condition, including the loss of humanist unity, is just the building block for the next phase of becoming subjects together."

39. In German, *der Sinn der Sache,* to which Nietzsche's wording alludes here, is idiomatic for purpose.

40. Geoff Waite notes that, in "our current postmodernist climate, all this sounds familiar, even *natural*" (*Nietzsche's Corps/E,* 278) and subsequently aligns Nietzsche with anyone from Benjamin, Foucault, Deleuze, Baudrillard, Derrida, and Lyotard to Eco, Austin, and even Gadamer.

41. Fuchs, "Circularity of the Embodied Mind," 7.

42. Stegmaier distinguishes Nietzsche's will to power, which represents a situative concept of power, from the concept of organized power in Luhmann's systems theory and that of latent power in Foucault's archaeology of power *dispositifs* (*Orientierung im Nihilismus,* 272–301 [ch. 9: "Superior Orientation: Nietzsche's, Luhmann's and Foucault's Demoralization of Power and The Contexts of Democracy"]).

43. Stigmergy provides "a general mechanism that relates individual and colony-level behaviors: individual behavior modifies the environment, which in turn modifies the behavior of other individuals.... Stigmergy is often associated with flexibility: when the environment changes because of an external perturbation, the insects respond appropriately to that perturbation, as if it were a modification of the environment caused by the colony's activities. In other words, the colony can collectively respond to the perturbation with individuals exhibiting the same behavior" (Bonabeau, Dorigo, and Theraulaz, *Swarm Intelligence,* 16).

44. Hayles, "Ethics for Cognitive Assemblages," 1201.

45. See Amoore, *Cloud Ethics,* for a more extensive analysis not only of the manifest social and political harm today's algorithmic data mining can cause but also of how "algorithms are generating the bounded conditions of what a democracy, a border crossing, a social movement, an election, or a public protest could be in the world" (4).

46. Richardson, *Nietzsche's Values,* 206. Furthermore, Richardson notes that, for Nietzsche, "morality is the way that herd-instinct still expresses itself in members who now think of themselves as (agential, *übersittlich,* autonomous) individuals" (230).

47. Lemm, "Truth, Embodiment, and Probity," 297.

48. Miyasaki, "Nietzschean Case for Illiberal Egalitarianism," 168.

49. Miyasaki, 167.

50. I am borrowing this definition of authenticity from David Wellbery, who sees it as the expression of a historical problematic that is also a core of Hegel's idealism and that finds its cognate poetic articulation in Goethe's postclassical work: "To live oneself, then, means that one lives *entirely* in one's doings, that one is fully in those doings and, for this reason, fully with one's own actions, willed by one's self, and in their unity they constitute one's self-lived life" ("Imagination of Freedom," 220). Nietzsche's admiration for Goethe is well known.

51. Baecker, "Hitler Swarm," 70–71.

52. Baecker, 71.

53. Baecker, 71.

54. Baecker, 71.

55. Baecker, 77.

56. As Harald Welzer has shown, the chilling historical evidence shows that, on an individual basis, moral codes did not simply erode or were overcome by the perpetrators (though viewed from the outside we can certainly argue so), but were substituted with "particular moral commitments and principles [that] gave the perpetrators a sense of moral integrity which enabled them to carry out the deeds they performed" ("Mass Murder and Moral Code," 30).

57. Braidotti, "Posthuman Critical Theory," 29.

58. Rorty, *Contingency, Irony, and Solidarity*, xv.

59. Luhmann, a sociologist, is aware of this quandary, acknowledging it most prominently in the title of his 1988 magnum opus, *Die Gesellschaft der Gesellschaft*.

60. See Stegmaier, *Orientierung im Nihilismus*, 267.

61. Latour, "Morality and Technology," 254.

62. Latour, 258.

63. Sharon argues along these lines: "We might imagine, for example, a guideline based on the idea that the technological expertise of a consumer technology company is needed to advance research (industry), but that GHRtype collaborations require a separation of roles (civic), whereby corporate actors cannot both provide the technical infrastructure used in the research project (i.e. collect and store data) and be in charge of data analysis. Another combinatory solution could entail that corporate actors are entitled to a return on investment (market), but that a tax on data use generated in the public domain (industry) redistributes wealth for public needs in the health sector (civic)" ("When Digital Health Meets Digital Capitalism," 9).

Bibliography

Abel, Günter. "Bewusstsein—Sprache—Natur: Nietzsches Philosophie des Geistes." *Nietzsche Studien* 30, no. 1 (2001): 1–43.
Abel, Günter. *Nietzsche: Die Dynamik der Willen zur Macht und die ewige Wiederkehr.* 2nd ed. Berlin: Walter de Gruyter, 1998.
Acampora, Christa Davis, and Ralph R. Acampora, eds. *A Nietzschean Bestiary: Becoming Animal beyond Docile and Brutal.* Lanham, Md.: Rowman & Littlefield, 2004.
Adorno, Theodor. *The Jargon of Authenticity.* Translated by Knut Tarnowski and Frederic Will. Evanston, Ill.: Northwestern University Press, 1973.
Althusser, Louis. "Marxism and Humanism." In *For Marx,* translated by Ben Brewster, 219–48. London: Allen Lane, 1969; repr. London: Verso, 2009. marxists.org/reference/archive/althusser/1964/marxism-humanism.htm.
Amoore, Louise A. *Cloud Ethics: Algorithms and the Attributes of Ourselves and Others.* Durham, N.C.: Duke University Press, 2020.
Ansell-Pearson, Keith. *Viroid Life: Perspectives on Nietzsche and the Transhuman Condition.* New York: Routledge, 1997.
Babich, Babette. "Towards Nietzsche's 'Critical' Theory—Science, Art, Life and Creative Economics." In *Nietzsche als Kritiker und Denker der Transformation,* edited by Helmut Heit and Sigridur Thorgeirsdottir, 112–33. Berlin: Walter de Gruyter, 2016.
Babich, Babette. "Friedrich Nietzsche and the Posthuman/Transhuman in Film and Television." In *The Palgrave Handbook of Posthumanism in Film and Television,* edited by Michael Hauskeller, Thomas D. Philbeck, and Curtis D. Carbonell, 45–53. New York: Palgrave Macmillan, 2015.
Baecker, Dirk. "The Hitler Swarm." *Thesis Eleven: Critical Theory and Historical Sociology* 117 (2013): 68–88.
Baecker, Dirk. *Wozu Kultur?* Berlin: Kulturverlag Kadmos, 2000.
Barad, Karen. *Meeting the Universe Halfway: Quantum Physics and the Entanglement of Matter and Meaning.* Durham, N.C.: Duke University Press, 2007.
Bateson, Gregory. "Form, Substance, and Difference." In *Steps to an Ecology of*

Mind: Collected Essays in Anthropology, Psychiatry, Evolution, and Epistemology, 455–73. Northvale, N.J.: Jason Aronson, 1972.

Beard, Charles A. "Time, Technology, and the Creative Spirit in Political Science." *American Political Science Review* 21, no. 1 (1927): 1–11.

Beckmann, Johann. *Entwurf einer allgemeinen Technologie*. Göttingen: J. F. Röwer, 1806.

Beckmann, Johann. *Anleitung zur Technologie oder zur Kenntniß der Handwerke, Fabriken und Manufacturen: vornehmlich derer, die mit der Landwirthschaft, Polizey und Cameralwissenschaft in nächster Verbindung stehn; nebst Beyträgen zur Kunstgeschichte*. Göttingen: Vandenhoeck & Ruprecht, 1777.

Beiner, Ronald. *Dangerous Minds: Nietzsche, Heidegger, and the Return of the Far Right*. Philadelphia: University of Pennsylvania Press, 2018.

Bennett, Jane. *Vibrant Matter: A Political Ecology of Things*. Durham, N.C.: Duke University Press, 2010.

Bertino, Christian Andrea. "Sprache und Instinkt bei Herder und Nietzsche." *Nietzsche Studien* 39, no. 1 (2010): 70–99.

Bonabeau, Eric, Marco Dorigo, and Guy Theraulaz. *Swarm Intelligence: From Natural to Artificial Systems*. Oxford: Oxford University Press, 1999.

Bostrom, Nick, et al. "Transhumanist FAQ." 2001–. humanityplus.org/transhumanist-faq.

Bostrom, Nick. "A History of Posthumanist Thought." *Journal of Evolution and Technology* 14, no. 1 (2005): 1–25.

Braidotti, Rosi. "A Theoretical Framework for the Critical Posthumanities." *Theory, Culture & Society* 36, no. 6 (2019): 31–61.

Braidotti, Rosi. "Posthuman Critical Theory." In *Critical Posthumanism and Planetary Futures*, edited by Debashish Banerji and Makarand R. Paranjape, 13–32. New Delhi: Springer, 2016.

Braidotti, Rosi. *The Posthuman*. Cambridge: Polity, 2013.

Braidotti, Rosi, and Maria Hlavajova, eds. *Posthuman Glossary*. London: Bloomsbury Academic, 2019.

Brobjer, Thomas H. "Nietzsche's Reading and Knowledge of Natural Science: An Overview." In *Nietzsche and Science*, edited by Gregory Moore and Thomas H. Brobjer, 21–50. Burlington, Vt.: Ashgate, 2004.

Brobjer, Thomas H. "Nietzsche's Knowledge of Marx and Marxism." *Nietzsche Studien* 31, no. 1 (2002): 298–313.

Bull, Malcolm. *Anti-Nietzsche*. London: Verso, 2011.

Butler, Judith. "Foucault and the Paradox of Bodily Inscription." *The Journal of Philosophy* 86, no. 11 (1989): 601–7.

Butler, Judith. "Performative Acts and Gender Constitution: An Essay in Phenomenology and Feminist Theory." *Theatre Journal* 40, no. 4 (1988): 519–31.

Cheung, Chiu-yee, and Adrian Hsia. "Nietzsche's Reception of Chinese Culture." *Nietzsche Studien* 32, no.1 (2003): 296–312.

Clark, Maudemarie. *Nietzsche on Truth and Philosophy*. Cambridge: Cambridge University Press, 1990.

Clark, Maudemarie, and David Dudrick. "Nietzsche on the Will: An Analysis of BGE 19." In *Nietzsche on Freedom and Autonomy*, edited by Ken Gemes and Simon May, 247–68. New York: Oxford University Press, 2009.

Clarke, Bruce. *Gaian Systems: Lynn Margulis, Neocybernetics, and the End of the Anthropocene*. Minneapolis: University of Minnesota Press, 2020.

Clarke, Bruce, and Manuela Rossini, eds. *The Cambridge Companion to Literature and the Posthuman*. Cambridge: Cambridge University Press, 2016.

Colebrook, Claire. *Death of the PostHuman: Essays on Extinction Vol. 1*. Ann Arbor, Mich.: Open Humanities, 2014.

Cowan, Michael. *Cult of the Will: Nervousness and German Modernity*. University Park: Pennsylvania State University Press, 2008.

Cowan, Michael. "'Nichts ist so sehr zeitgemäss als Willensschwäche': Nietzsche and the Psychology of the Will." *Nietzsche Studien* 34, no. 1 (2005): 48–74.

Crary, Jonathan. *Techniques of the Observer: On Vision and Modernity in the Nineteenth Century*. Cambridge, Mass.: MIT Press, 1990.

Crawford, Claudia. *The Beginnings of Nietzsche's Theory of Language*. Berlin: Walter de Gruyter, 1988.

Darwin, Charles. *On the Origin of Species*. Hazleton: Pennsylvania State University Press, 2001.

Deleuze, Gilles, and Félix Guattari. *A Thousand Plateaus*. Translated and introduced by Brian Massumi. Vol. 2 of *Capitalism and Schizophrenia*. London: Bloomsbury, 2015.

Derrida, Jacques. *Spurs: Nietzsche's Styles. Éperons: Les Styles de Nietzsche*. Translated by Barbara Harlow. Chicago: University of Chicago Press, 1973.

Derrida, Jacques, and Richard Beardsworth. "Nietzsche and the Machine." *Journal of Nietzsche Studies* 7 (1994): 7–66.

Dries, Manuel. "Early Nietzsche on History, Embodiment and Value." In *Nietzsche on Consciousness and the Embodied Mind*, edited by Manuel Dries, 49–70. Berlin: Walter de Gruyter, 2018.

Drochon, Hugo. "'An Old Carriage with New Horses': Nietzsche's Critique of Democracy." *History of European Ideas* 42, no. 8 (2016): 1055–68.

Dünkelberg, Friedrich Wilhelm. *Encyclopädie und Methodologie der Kulturtechnik. Zum Gebrauche an landwirthschaftlichen und technischen Lehranstalten und zum Selbstunterricht für Landwirte, Techniker und Verwaltungsbeamte.* Braunschweig: Friedrich Bieweg and Son, 1883.

Dunkle, Ian D. "Moral Physiology and Vivisection of the Soul: Why Does Nietzsche Criticize the Life Sciences?" *Inquiry* 61, no.1 (2018): 62–81.

Elias, Norbert. *The Civilizing Process: Sociogenetic and Psychogenetic Investigations.* Rev. ed. Translated by Edmund Jephcott, edited by Eric Dunning, Johan Goudsblom, and Stephen Mennell. Oxford: Blackwell, 2000.

Emden, Christian J. "Histories of Violence: Nietzsche on Cruelty and Normative Order." In *Nietzsche on Memory and History,* edited by Anthony K. Jensen and Carlotta Santini, 209–37. Berlin: Walter de Gruyter, 2021.

Emden, Christian J. "Agency without Humans: Normativity and Path Dependence in the Nineteenth-Century Life Sciences." In Landgraf, Trop, and Weatherby, *Posthumanism in the Age of Humanism,* 53–77.

Emden, Christian J. "Nietzsche's Will to Power: Biology, Naturalism, and Normativity." *The Journal of Nietzsche Studies* 47, no. 1 (2016): 30–60.

Emden, Christian J. *Nietzsche's Naturalism: Philosophy and the Life Sciences in the Nineteenth Century.* Cambridge: Cambridge University Press, 2014.

Emden, Christian J. *Nietzsche on Language, Consciousness, and the Body.* Urbana: University of Illinois Press, 2005.

Engler-Coldren, Katharina, Lore Knapp, and Charlotte Lee. "Embodied Cognition around 1800: Introduction." *German Life and Letters* 70, no. 4 (2017): 413–22.

Espinas, Alfred. *Die thierischen Gesellschaften: eine vergleichend-psychologische Untersuchung.* Braunschweig: Friedrich Bieweg and Son, 1879.

Esposito, Roberto. *Bios: Biopolitics and Philosophy.* Translated and with an introduction by Timothy Campbell. Minneapolis: University of Minnesota Press, 2008.

Ferrando, Francesca. *Philosophical Posthumanism.* London: Bloomsbury Academic, 2019.

Fornari, Maria Cristina. "Social Ties and the Emergence of the Individual: Nietzsche and the English Perspective." In *Nietzsche and the Becoming of Life,* edited by Vanessa Lemm, 234–53. New York: Fordham University Press, 2014.

Fornari, Maria Cristina. "Die Spur Spencers in Nietzsches 'Moralischem Bergwerke.'" *Nietzsche Studien* 34, no.1 (2008): 310–28.

Foucault, Michel. "What Is Enlightenment?" In *The Foucault Reader,* edited by Paul Rabinow, 32–50. New York: Penguin, 1984.

Foucault, Michel. *Discipline and Punish: The Birth of the Prison.* Translated by Alan Sheridan. New York: Random, 1979.

Foucault, Michel. *The Order of Things: An Archeology of the Human Sciences.* New York: Pantheon, 1971.

Fuchs, Thomas. "The Circularity of the Embodied Mind." *Frontiers in Psychology* 11 (2020): 1–13.

Fuller, Steve. *Nietzschean Meditations.* Posthuman Studies 1. Basel: Schwabe, 2020.

Galindo, Martha Zapata. *Triumph des Willens zur Macht: zur Nietzsche-Rezeption im NS-Staat.* Hamburg: Argument, 1995.

Gerber, Gustav. *Die Sprache als Kunst.* Bromberg: Mittlersche Buchhandlung, 1871.

Gordon, Deborah. *Ant Encounters: Interaction Networks and Colony Behavior.* Princeton, N.J.: Princeton University Press, 2011.

Graham, Elaine. "'Nietzsche Gets a Modem': Transhumanism and the Technological Sublime." *Literature and Theology* 16, no. 1 (2002): 65–80.

Grimm, Jacob and Wilhelm, eds. *Deutsches Wörterbuch.* 16 vols. Leipzig: S. Hirzel, 1854–1961.

Guerreschi, Luca. "Leib, Seele und Subjektivität nach Nietzsche: internationale Perspektiven auf ein Problem im Wandel." *Nietzsche Studien* 50, no. 1 (2021): 340–60.

Habermas, Jürgen. *The Philosophical Discourse of Modernity: Twelve Lectures.* Translated by Frederick Lawrence. Cambridge: Polity, 1990.

Hamacher, Werner. "'Disgregation des Willens'—Nietzsche über Individuum und Individualität." *Nietzsche Studien* 15, no. 1 (1986): 306–36.

Haraway, Donna J. *Staying with the Trouble: Making Kin in the Chthulucene.* Durham, N.C.: Duke University Press, 2016.

Haraway, Donna J. *When Species Meet.* Minneapolis: University of Minnesota Press, 2008.

Haraway, Donna J. "A Cyborg Manifesto: Science, Technology, and Socialist-Feminism in the Late Twentieth Century." In *Simians, Cyborgs, and Women: The Reinvention of Nature,* 149–82. New York: Routledge, 1991.

Hartmann, Eduard von. *Philosophie des Unbewussten.* 10th ed. Leipzig: Wilhelm Friedrich, 1890.

Hatab, Lawrence J. "Talking Ourselves into Selfhood: Nietzsche on Consciousness and Language in Gay Science 354." In *Nietzsche on Consciousness and the Embodied Mind,* edited by Manuel Dries, 183–94. Berlin: Walter de Gruyter, 2018.

Hatab, Lawrence J. "Breaking the Contract Theory: The Individual and the Law in Nietzsche's Genealogy." In Siemens and Roodt, *Nietzsche, Power and Politics,* 169–88.

Hauskeller, Michael, Thomas D. Philbeck, and Curtis D. Carbonell. "Posthumanism in Film and Television." In *The Palgrave Handbook of Posthumanism in Film and Television,* edited by Michael Hauskeller, Thomas D. Philbeck, and Curtis D. Carbonell, 1–7. New York: Palgrave McMillan, 2015.

Hayles, N. Katherine. "Ethics for Cognitive Assemblages: Who's in Charge Here?" In Callus et. al, *Palgrave Handbook of Critical Posthumanism,* 1195–224.

Hayles, N. Katherine. *Unthought: The Power of the Cognitive Nonconscious.* Chicago: University of Chicago Press, 2017.

Hayles, N. Katherine. "The Cognitive Nonconscious and the New Materialisms." In *The New Politics of Materialism: History, Philosophy, Science,* edited by Sarah Ellenzweig and John H. Zammito, 181–99. New York: Routledge, 2017.

Hayles, N. Katherine. *How We Became Posthuman: Virtual Bodies in Cybernetics, Literature, and Informatics.* Chicago: University of Chicago Press, 1999.

Heidegger, Martin. *Basic Writings: From "Being and Time" (1927) to "The Task of Thinking" (1964).* Rev. ed. Edited by David Farrell Krell. New York: Harper & Row, 1993.

Heidegger, Martin. *Platons Lehre von der Wahrheit, mit einem Brief über den Humanismus.* 2nd ed. Bern: Francke, 1954.

Heidelberger, Michael. "Beziehungen zwischen Sinnesphysiologie und Philosophie im 19. Jahrhundert." In *Philosophie und Wissenschaften: Formen und Prozesse ihrer Interaktion,* edited by Hans Jörg Sandkühler, 37–58. Frankfurt am Main: Peter Lang, 1997.

Heit, Helmut. "Naturalizing Perspectives: On the Epistemology of Nietzsche's Experimental Naturalizations." *Nietzsche Studien* 45, no. 1 (2016): 56–80.

Helmholtz, Hermann von. *Handbuch der physiologischen Optik.* Leipzig: L. Voss, 1867.

Helmholtz, Hermann von. *Die Lehre von der Tonempfindung als physiologische Grundlage für die Theorie der Musik.* 4th ed. Braunschweig: Friedrich Bieweg and Son, 1877.

Herbrechter, Stefan. *Posthumanism: A Critical Analysis.* London: Bloomsbury, 2013.

Herbrechter, Stefan, and Ivan Callus. "Introduction—Shakespeare Ever After." In *Posthumanist Shakespeares,* edited by Stefan Herbrechter and Ivan Callus, 1–19. New York: Palgrave McMillan, 2012.

Herbrechter, Stefan, Ivan Callus, Manuela Rossini, Marija Grech, et al., eds. *The Palgrave Handbook of Critical Posthumanism*. Cham, Switzerland: Palgrave Macmillan, 2022.

Holland, Jocelyn. "Instruction in an Imperfect Science: Challenges in Defining—and Teaching—Technology around 1800." *Amodern* 9. April 2020. amodern.net/article/imperfect-science/.

Holland, Jocelyn. "Angeln, Blatt, Constellation: Plural Forms in Nietzsche's *Ueber Wahrheit und Lüge im aussermoralischen Sinne*." *MLN* 126 (2011): 518–33.

Holland, Jocelyn, and Edgar Landgraf, eds. *The Archimedean Point in Modernity*. Special issue of *SubStance* 43, no. 3 (2014).

Holub, Robert C. *Nietzsche in the Nineteenth Century: Social Questions and Philosophical Interventions*. Philadelphia: University of Philadelphia Press, 2018.

Holub, Robert C. *Nietzsche's Jewish Problem: Between Anti-Semitism and Anti-Judaism*. Princeton, N.J.: Princeton University Press, 2015.

Hörl, Erich. "Introduction to General Ecology: The Ecologization of Thinking." Translated by Nils F. Schott. In *General Ecology: The New Ecological Paradigm*, edited by Erich Hörl with James Burton, 1–73. London: Bloomsbury, 2017.

Huddleston, Andrew. "'Consecration to Culture': Nietzsche on Slavery and Human Dignity." *Journal of the History of Philosophy* 52, no. 1 (2014): 135–60.

Humboldt, William von. *One Language: On the Diversity of Human Language Construction and Its Influence on the Mental Development of the Human Species*. Translated by Peter Heath. Edited by Michael Losonsky. Cambridge: Cambridge University Press, 1999.

Janaway, Christopher. *Schopenhauer: A Very Short Introduction*. Oxford: Oxford University Press, 2002.

Janaway, Christopher, and Simon Robertson, eds. *Nietzsche, Naturalism and Normativity*. Oxford: Oxford University Press, 2012.

Johach, Eva. "Der Bienenstaat: Geschichte eines politisch-moralischen Exempels." In *Politische Zoologie*, edited by Anne von der Heiden and Joseph Vogl, 219–33. Berlin: Diaphanes, 2007.

Johnson, Steven. *Emergence: The Connected Lives of Ants, Brains, Cities, and Software*. New York: Scribner, 2001.

Kant, Immanuel. *Critique of Judgment*. Translated by Werner S. Pluhar. Indianapolis, Ind.: Hackett, 1987.

Kapp, Ernst. *Elements of a Philosophy of Technology: On the Evolutionary History of Culture*. Edited and introduced by Jeffrey W. Kirkwood, and Leif Weatherby. Translated by Lauren K. Wolfe. Minneapolis: University of Minnesota Press, 2018.

Katsafanas, Paul. "Philosophical Psychology as a Basis for Ethics." *The Journal of Nietzsche Studies* 44, no. 2 (2013): 297–314.

Kim, Alan. "Wilhelm Maximilian Wundt." In *The Stanford Encyclopedia of Philosophy*, edited by Edward N. Zalta. June 2016; revised September 2016. plato.stanford.edu/archives/fall2016/entries/wilhelm-wundt/.

Kittler, Friedrich A. "Nietzsche (1844–1900)." In *The Truth of the Technological World: Essays on the Genealogy of Presence*, 17–30. Stanford, Calif.: Stanford University Press, 2014.

Kittler, Friedrich A. *Discourse Networks 1800/1900*. Translated by Michael Metteer, with Chris Cullens. Foreword by David E. Wellbery. Stanford, Calif.: Stanford University Press, 1990.

Kroker, Arthur. *The Will to Technology and the Culture of Nihilism: Heidegger, Nietzsche, and Marx*. Toronto: University of Toronto Press, 2004.

Kunnas, Tarmo. "Nietzsche und die Technologie." *Text & Kontext* 13, no. 2 (1985): 315–19.

Kurzweil, Ray. *The Singularity Is Near: When Humans Transcend Biology*. New York: Penguin, 2005.

Lambrecht, Georg Friedrich von. *Lehrbuch der Technologie oder Anleitung zur Kenntniß der Handwerke, Fabriken und Manufakturen*. Halle: Hammerdesche Buchhandlung, 1787.

Landgraf, Edgar. "Embodied Phantasy: Johannes Müller and the Nineteenth-Century Neurophysiological Foundations of Critical Posthumanism." In Landgraf, Trop, and Weatherby, *Posthumanism in the Age of Humanism*, 79–101.

Landgraf, Edgar. "Improvisation, Posthumanism, and Agency in Art (Gerhard Richter Painting)." *Liminalities: A Journal of Performance Studies* 14, no. 1 (2018): 207–22.

Landgraf, Edgar. "Circling the Archimedean Viewpoint: Observations of Physiology in Nietzsche and Luhmann." *The Archimedean Point in Modernity*, edited by Jocelyn Holland and Edgar Landgraf. Special issue of *SubStance* 43, no. 3 (2014): 88–106.

Landgraf, Edgar. "The Physiology of Observation in Nietzsche and Luhmann." *Monatshefte* 105, no. 3 (Fall 2013): 472–88.

Landgraf, Edgar. *Improvisation as Art. Conceptual Challenges, Historical Perspectives*. New York: Continuum, 2011.

Landgraf, Edgar. "The Disintegration of Modern Culture: Nietzsche and the Information Age." *Comparative Literature* 57 (2005): 25–44.
Landgraf, Edgar, Gabriel Trop, and Leif Weatherby, eds. *Posthumanism in the Age of Humanism: Mind, Matter, and the Life Sciences After Kant.* London: Bloomsbury, 2019.
Lange, Friedrich Albert. *Geschichte des Materialismus und Kritik seiner Bedeutung in der Gegenwart.* Leipzig: J. Baedeker, 1887.
Large, Duncan. "Nietzsche, Burckhardt, and the Concept of Culture (from the North American Nietzsche Society Papers at the World Congress of Philosophy, August 1998—'Our Greatest Teacher')." *International Studies in Philosophy* 32, no. 3 (2000): 3–25.
Latour, Bruno. *Reassembling the Social: An Introduction to Actor-Network-Theory.* Oxford: Oxford University Press, 2005.
Latour, Bruno. "Morality and Technology: The End of the Means." Translated by Couze Venn. *Theory, Culture & Society* 19, no. 5–6 (2002): 247–60.
Latour, Bruno. "On Recalling ANT." In *Actor Network and After,* edited by John Law and John Hassard, 15–25. Oxford: Blackwell, 1999.
Latour, Bruno. "Technology is Society Made Durable." *The Sociological Review* 38, suppl. 1 (1990): 103–31.
Leiber, Theodor. *Vom mechanistischen Weltbild zur Selbstorganisation des Lebens: Helmholtz' und Boltzmanns Forschungsprogramme und ihre Bedeutung für Physik, Chemie, Biologie und Philosophie.* Freiburg: Karl Alber, 2000.
Leitner, Brian. "Nietzsche's Hatred of 'Jew Hatred.'" Review of *Nietzsche's Jewish Problem: Between Anti-Semitism and Anti-Judaism* by Robert C. Holub. *The New Rambler.* December 21, 2015. newramblerreview.com/book-reviews/philosophy/nietzsche-s-hatred-of-jew-hatred.
Lemm, Vanessa. *Homo Natura: Nietzsche, Philosophical Anthropology and Biopolitics.* Edinburgh: Edinburgh University Press, 2020.
Lemm, Vanessa. "Truth, Embodiment, and Probity (*Redlichkeit*) in Nietzsche." In *Nietzsche on Consciousness and the Embodied Mind,* edited by Manuel Dries, 289–307. Berlin: Walter de Gruyter, 2018.
Lemm, Vanessa. "Shedding New Light on the Ethical Dimension of Nietzsche's Philosophy." *Nietzsche Studien* 43, no. 1 (2014): 347–58.
Lemm, Vanessa. *Nietzsche's Animal Philosophy: Culture, Politics, and the Animality of the Human Being.* New York: Fordham University Press, 2009.
Lende, Daniel H., and Greg Downey, eds. *The Encultured Brain: An Introduction to Neuroanthropology.* Cambridge, Mass.: MIT Press, 2012.

Lotze, Hermann. *Allgemeine Pathologie und Therapie als mechanische Naturwissenschaften.* Leipzig: Weidmann, 1842.
Luhmann, Niklas. *Art as a Social System.* Translated by Eva M. Knodt. Stanford, Calif.: Stanford University Press, 2000.
Luhmann, Niklas. *Die Gesellschaft der Gesellschaft.* Frankfurt am Main: Suhrkamp, 1998.
Luhmann, Niklas. "Politische Steuerungsfähigkeit eines Gemeinwesens." In *Die Gesellschaft für morgen,* edited by Reinhard Göhner, 50–65. Munich: Piper, 1993.
Luhmann, Niklas. *Die Wissenschaft der Gesellschaft.* Frankfurt am Main: Suhrkamp, 1990.
Luhmann, Niklas. "Technology, Environment and Social Risk: A Systems Perspective." In *Technological Risk and Political Conflict: Perspectives from West Germany.* Special issue of *Industrial Crisis Quarterly* 4, no. 3 (1990): 223–31.
Lyotard, Jean-François. *The Inhuman: Reflections on Time.* Translated by Geoffrey Bennington and Rachel Bowlby. Stanford, Calif.: Stanford University Press, 1991.
Malafouris, Lambros. *How Things Shape the Mind: A Theory of Material Engagement.* Cambridge, Mass.: MIT Press, 2013.
Marx, Karl. *Capital: Volume 1.* Marx and Engels Collected Works 35. London: Lawrence and Wishart, 1936.
Marx, Karl. *Grundrisse: Foundations of the Critique of Political Economy.* London: Penguin, 1993.
Marx, Karl. "Economic and Philosophical Manuscripts (1844)." In *Early Writings,* translated by Rodney Livingstone and Gregor Benton, 279–400. New York: Penguin Classics, 1992.
Marx, Karl. *Das Kapital: Kritik der politischen Oekonomie.* Hamburg: Otto Meissner, 1867.
Marx, Karl, and Frederick Engels. *The Manifesto of the Communist Party.* 1848. Translated by Samuel Moor in cooperation with Frederick Engels. In vol. 1 of *Marx/Engels Selected Works,* 98–137. Moscow: Progress, 1969. Edited by Andy Blunden for marxists.org/archive/marx/works/1848/communist-manifesto/.
Marx, Leo. "Technology: The Emergence of a Hazardous Concept." *Technology and Culture* 51, no. 3 (2010): 561–77.
Massumi, Brian. "The Supernormal Animal." In *The Nonhuman Turn,* edited by Richard Grusin, 1–17. Minneapolis: University of Minnesota Press, 2015.

Maturana, Humberto, and Francisco Varela. *Autopoiesis and Cognition: The Realization of the Living.* Dordrecht, Holland: D. Reidel, 1980.

MacCormack, Patricia. *Posthuman Ethics.* New York: Routledge, 2012.

Meillassoux, Quentin. *After Finitude: An Essay on the Necessity of Contingency.* Translated by Ray Brassier. London: Bloomsbury, 2010.

Mellamphy, Dan, and Nandita Biswas Mellamphy. "Nietzsche and Networks, Nietzschean Networks: The Digital Dionysus." In *The Digital Dionysus: Nietzsche and the Network-Centric Condition,* edited by Dan and Nandita Biswas Mellamphy, 10–30. New York: Punctum, 2016.

Meredith, Thomas R. "Bound Sovereignty: The Origins of Moral Conscience in Nietzsche's 'Sovereign Individual.'" *Nietzsche Studien* 50, no. 1 (2021): 217–43.

Miyasaki, Donovan. "A Nietzschean Case for Illiberal Egalitarianism." In *Nietzsche as Political Philosopher,* edited by Manuel Knoll and Barry Stocker, 155–70. Berlin: Walter de Gruyter, 2014.

Moore, Gregory. "Introduction." In *Nietzsche and Science,* edited by Gregory Moore and Thomas H. Brobjer, 1–17. Burlington, Vt.: Ashgate, 2004.

More, Max. "The Overhuman in the Transhuman." *Journal of Evolution and Technology* 21, no. 1 (2010): 1–4.

Morton, Timothy. *Hyperobjects: Philosophy and Ecology after the End of the World.* Minneapolis: University of Minnesota Press, 2013.

Mouffe, Chantal. *On the Political.* New York: Routledge, 2005.

Müller, Jan-Werner. *What is Populism?* Philadelphia: University of Pennsylvania Press, 2016.

Müller, Johannes. *Handbuch der Physiologie des Menschen für Vorlesungen.* Koblenz: J. Hölscher, 1840–1844.

Müller, Johannes. *Über die phantastischen Gesichtserscheinungen.* Koblenz: Hölscher, 1826.

Müller, Johannes. *Zur vergleichenden Physiologie des Gesichtssinnes des Menschen und der Thiere nebst einem Versuch über die Bewegungen der Augen und über den menschlichen Blick.* Leipzig: G. Gnobloch, 1826.

Nancy, Jean-Luc. *The Banality of Heidegger.* Translated by Jeff Fort. New York: Fordham University Press, 2017.

Nayar, Pramod K. *Posthumanism.* Cambridge: Polity, 2014.

Nietzsche, Friedrich. *On the Genealogy of Morals* (GM). Edited by Keith Ansell-Pearson. Translated by Carol Diethe. Cambridge: Cambridge University Press, 2006.

Nietzsche, Friedrich. *Beyond Good and Evil: Prelude to a Philosophy of the Future*

(BGE). Translated by Judith Norman. Cambridge: Cambridge University Press, 2002.

Nietzsche, Friedrich. *The Gay Science* (GS). Translated by Josefine Nauckoff. Cambridge: Cambridge University Press, 2001.

Nietzsche, Friedrich. *The Birth of Tragedy* (BT). Translated by Douglas Smith. Oxford: Oxford University Press, 2000.

Nietzsche, Friedrich. *Twilight of the Idols, or How to Philosophize with the Hammer* (TI). Translated by Richard Polt. Indianapolis, Ind.: Hackett, 1997.

Nietzsche, Friedrich. *Daybreak: Thoughts on the Prejudices of Morality* (D). Translated by R. J. Hollingdale. Cambridge: Cambridge University Press, 1997.

Nietzsche, Friedrich. *Human, All Too Human* (HH). Translated by R. J. Hollingdale. Cambridge: Cambridge University Press, 1996.

Nietzsche, Friedrich. *On the Uses and Disadvantages of History for Life* (HL). In *Untimely Meditations,* edited by Daniel Breazeale, translated by R. J. Hollingdale, 57–124. Cambridge: Cambridge University Press, 1997.

Nietzsche, Friedrich. *Sämtliche Briefe: Kritische Studienausgabe Briefe* (KSB). Edited by Giorgio Colli and Mazzino Montinari. 8 vols. Munich: Walter de Gruyter, 1986.

Nietzsche, Friedrich. *Sämtliche Werke: Kritische Studienausgabe* (KSA). Edited by Giorgio Colli eand Mazzino Montinari. 15 vols. Munich: Walter de Gruyter, 1980.

Nietzsche, Friedrich. *On Truth and Lies in a Nonmoral Sense* (TL). In *Philosophy and Truth: Selections from Nietzsche's Notebooks of the Early 1870's,* edited and translated by Daniel Breazeale, 79–97. Atlantic Highlands, N.J.: Humanities Press, 1979.

Nietzsche, Friedrich. *Werke: Kritische Gesamtausgabe* (KGW). Edited by Giorgio Colli and Mazzino Montinari. Munich: Walter de Gruyter, 1967.

Nietzsche, Friedrich. *The Antichrist* (A). Translated by Walter Kaufmann. New York: Penguin, 1954.

Otis, Laura. *Networking: Communicating with Bodies and Machines in the Nineteenth Century.* Ann Arbor: University of Michigan Press, 2011.

Otis, Laura. *Müller's Lab: The Story of Jakob Henle, Theodor Schwann, Emil du Bois-Reymond, Hermann von Helmholtz, Rudolf Virchow, Robert Remak, Ernst Haeckel, and Their Brilliant, Tormented Advisor.* Oxford: Oxford University Press, 2007.

Owen, David. "Nietzsche, Ethical Agency and the Problem of Democracy." In Siemens and Roodt, *Nietzsche, Power and Politics,* 143–67.

Parikka, Jussi. *Insect Media: An Archaeology of Animals and Technology.* Minneapolis: University of Minnesota Press, 2010.
Parry, Richard. "*Episteme* and *Techne.*" In *The Stanford Encyclopedia of Philosophy,* edited by Edward N. Zalta. April 2003; revised March 2020. plato.stanford.edu/entries/episteme-techne/#6.
Patton, Paul. "Recent Work on Nietzsche's Social and Political Philosophy." *Nietzsche Studien* 50, no. 1 (2021): 382–95.
Patton, Paul. "Nietzsche on Rights, Power and the Feeling of Power." In Siemens and Roodt, *Nietzsche, Power and Politics,* 471–88.
Pepperell, Robert. *The Posthuman Condition: Consciousness Beyond the Brain.* Bristol, UK: Intellect Books, 2003.
Pippin, Robert B. *Nietzsche, Psychology, and First Philosophy.* Chicago: University of Chicago Press, 2010.
Pippin, Robert B. "Lightning and Flash, Agent and Deed (GM I: 6–17)." In *Friedrich Nietzsche's "On the Genealogy of Morals": Critical Essays,* edited by Christa Davis Acampora, 131–45. Lanham, Md.: Rowman & Littlefield, 2006.
Plessner, Helmuth. *Gesammelte Schriften,* edited by Günter Dux et al. 10 vols. Frankfurt am Main: Suhrkamp, 1980–1985.
Rasch, William. *Sovereignty and Its Discontents: On the Primacy of Conflict and the Structure of the Political.* London: Birkbeck Law Press, 2004.
Rebentisch, Juliane. *The Art of Freedom: On the Dialectics of Democratic Existence.* Translated by Joesph Ganahl. Cambridge: Polity, 2016.
Reuter, Sören. "Reiz—Bild—Unbewusste Anschauung." *Nietzsche Studien* 33, no. 1 (2004): 351–72.
Rhode, Klaus. "Instinkt." In *Historisches Wörterbuch der Philosophie,* edited by Joachim Ritter et al., 408–17. Basel: Schabe WBG, 1998.
Richardson, John. *Nietzsche's Values.* New York: Oxford University Press, 2020.
Roden, David. *Posthuman Life: Philosophy at the Edge of the Human.* London: Routledge, 2015.
Rorty, Richard. *Contingency, Irony, and Solidarity.* Cambridge: Cambridge University Press, 1989.
Schatzberg, Eric. "*Technik* Comes to America: Changing Meanings of Technology before 1930." *Technology and Culture* 47, no. 3 (2006): 486–512.
Schloegel, Judy Johns, and Henning Schmidgen. "General Physiology, Experimental Psychology, and Evolutionism: Unicellular Organisms as Objects of Psychophysiological Research, 1877–1918." *Isis* 93, no. 4 (2002): 614–45.

Schlüpmann, Heide. *Friedrich Nietzsches ästhetische Opposition: Der Zusammenhang von Sprache, Natur und Kultur in seinen Schriften 1869–1876.* Stuttgart: Metzlersche, 1977.

Schneider, Georg Heinrich. *Der thierische Wille: Systematische Darstellung und Erklärung der thierischen Triebe und deren Entstehung, Entwicklung und Verbreitung im Thierreiche als Grundlage zu einer vergleichenden Willenslehre.* Leipzig: Abel, 1880.

Schopenhauer, Arthur. *The World as Will and Idea.* Translated by R. B. Haldane and J. Kemp. Vol. 1. London: Kegan Paul, Trench, and Trübner, 1909.

Schrift, Alan D. "Arachnophobe or Arachnophile? Nietzsche and His Spiders." In Acampora and Acampora, *Nietzschean Bestiary,* 61–70.

Sharon, Tamar. "When Digital Health Meets Digital Capitalism: How Many Common Goods Are at Stake?" *Big Data & Society* 5, no. 2 (2018): 1–12.

Sharon, Tamar. *Human Nature in an Age of Biotechnology: The Case for Mediated Posthumanism.* Philosophy of Engineering and Technology 14. New York: Springer, 2014.

Sharpe, Matthew. "Golden Calf: Deleuze's Nietzsche in the Time of Trump." *Thesis Eleven* 163, no. 1 (2021): 71–88.

Siegert, Bernhard. "Cultural Techniques, or the End of the Intellectual Postwar Era in German Media Theory." *Theory, Culture & Society* 30, no. 6 (2013): 48–65.

Siemens, Herman, and James Pearson, eds. *Conflict and Contest in Nietzsche's Philosophy.* London: Bloomsbury, 2019.

Siemens, Herman. "Nietzsche's Socio-Physiology of the Self." In *Nietzsche and the Problem of Subjectivity,* vol. 5 of *Nietzsche Today,* edited by João Constâncio, Maria João Mayer Branco, and Bartholomew Ryan, 629–53. Berlin: Walter de Gruyter, 2015.

Siemens, Herman, and Vasti Roodt, eds. *Nietzsche, Power and Politics: Rethinking Nietzsche's Legacy for Political Thought.* Berlin: Walter de Gruyter, 2008.

Simonin, David. "Nietzsches Lektüre von Alfred Espinas' Die thierischen Gesellschaften." In *Nietzsche als Leser,* edited by Hans-Peter Anschütz, Armin Thomas Müller, Mike Rottmann, and Yannick Souladié, 301–24. Berlin, Boston: Walter de Gruyter, 2021.

Sleigh, Charlotte. *Six Legs Better: A Cultural History of Myrmecology.* Baltimore, Md.: Johns Hopkins University Press, 2007.

Soper, Kate. "The Humanism in Posthumanism." *Comparative Critical Studies* 9, no. 3 (2012): 365–78.

Solies, Dirk. "Naturwissenschaften als Aufklärung? Am Beispiel von Nietzsches Physiologierezeption." In *Nietzsche als Radikalaufklärer oder radikaler Gegenaufklärer?*, edited by Renate Reschke, 247–54. Berlin: Akademie Verlag, 2004.

Sorgner, Stefan Lorenz. "Nietzsche als Ahnherr des Posthumanismus in den Künsten: Reflexionen zum Verhältnis von Bild, Wort und Ton." In *Bilder—Sprache—Künste: Nietzsches Denkfiguren im Zusammenhang*, edited by Renate Reschke, 45–58. Berlin: Akademie Verlag, 2011.

Sorgner, Stefan Lorenz. "Nietzsche, the Overman, and Transhumanism." *Journal of Evolution and Technology* 20, no.1 (2009): 29–42.

Spivak, Gayatri Chakravorty. *A Critique of Postcolonial Reason: Toward a History of the Vanishing Present*. Cambridge, Mass.: Harvard University Press, 1999.

Stegmaier, Werner. *Orientierung im Nihilismus—Luhmann meets Nietzsche*. Berlin: Walter de Gruyter, 2016.

Stiegler, Bernard. *Technics and Time: The Fault of Epimetheus*. Stanford, Calif.: Stanford University Press, 1998.

Stout, Dietrich, Nicholas Toth, Kathy Schick, and Thierry Chaminade. "Neural Correlates of Early Stone Age Toolmaking: Technology, Language and Cognition in Human Evolution." *Philosophical Transactions of the Royal Society of London, Series B, Biological Sciences* 363, no. 1499 (2008): 1939–49.

Thompson, Evan, and Francisco J. Varela. "Radical Embodiment: Neural Dynamics and Consciousness." *Trends in Cognitive Sciences* 5, no. 10 (2001): 418–25.

Thomsen, Mads Rosendahl, and Jacob Wamberg, eds. *The Bloomsbury Handbook of Posthumanism*. London: Bloomsbury Academic, 2020.

Tuncel, Yunus, ed. *Nietzsche and Transhumanism: Precursor or Enemy?* Newcastle, UK: Cambridge Scholars Publishing, 2017.

Uexküll, Jakob von. *Der Sinn des Lebens: Gedanken über die Aufgaben der Biologie mitgeteilt in einer Interpretation der zu Bonn 1824 gehaltenen Vorlesung des Johannes Müller "Von dem Bedürfnis der Physiologie nach einer philosophischen Naturbetrachtung."* Godesberg: Helmut Küpper, 1947.

Ure, Michael. "Resentment/Ressentiment." *Constellations* 22, no. 4 (2015): 599–613.

Vita-More, Natasha. "The Transhumanist Manifesto." 1993; 1998 (version 2); 2008 (version 3); 2020 (version 4). humanityplus.org/the-transhumanist-manifesto.

Waite, Geoff. *Nietzsche's Corps/E: Aesthetics, Politics, Prophecy, or, the Spectacular Technoculture of Everyday Life*. Durham, N.C.: Duke University Press, 1996.

Weatherby, Leif. "Farewell to Ontology: Hegel after Humanism." In Landgraf, Trop, and Weatherby, *Posthumanism in the Age of Humanism*, 145–64.
Weatherby, Leif. *Transplanting the Metaphysical Organ: German Romanticism between Leibniz and Marx*. New York: Fordham University Press, 2016.
Wellbery, David E. "The Imagination of Freedom: Goethe and Hegel as Contemporaries." In *Goethe's Ghosts: Reading and the Persistence of Literature*, edited by Simon Richter and Richard Block, 217–38. Rochester, N.Y.: Camden House, 2013.
Wellbery, David E. "Die Strategie des Paradoxons—Nietzsches Verhältnis zur Aufklärung." In *Aufklärung als Form: Beiträge zu einem historischen und aktuellen Problem*, edited by Helmut Schmiedt and Helmut J. Schneider, 161–72. Würzburg: Königshausen and Neumann, 1997.
Wellbery, David E. "Foreword." In Friedrich A. Kittler, *Discourse Networks 1800 / 1900*, translated by Michael Metteer, with Chris Cullens, vii–xxxiii. Stanford, Calif.: Stanford University Press, 1990.
Welzer, Harald. "Mass Murder and Moral Code: Some Thoughts on an Easily Misunderstood Subject." *History of the Human Sciences* 17, no. 2–3 (2004): 15–32.
Werber, Niels. "Schwärme, soziale Insekten, Selbstbeschreibungen der Gesellschaft: eine Ameisenfabel." In *Schwärme—Kollektive ohne Zentrum: eine Wissensgeschichte zwischen Leben und Information*, edited by Eva Horn and Lucas Marco Gisi, 183–202. Bielefeld: transcript, 2009.
Wilke, Tobias. "At the Intersection of Nervous System and Soul: Observation and Its Limits in Late 19th-Century Psychological Aesthetics." *Monatshefte* 105, no. 3 (2013): 443–57.
Williams-Hatala, Erin Marie, Kevin G. Hatala, McKenzie Gordon, Alastair Key, et al. "The Manual Pressures of Stone Tool Behaviors and Their Implications for the Evolution of the Human Hand." *Journal of Human Evolution* 119 (2018): 14–26.
Winthrop-Young, Geoffrey. "Cultural Studies and German Media Theory. " In *New Cultural Studies: Adventures in Theory*, edited by Gary Hall and Clave Birchall, 88–103. Athens: University of Georgia Press, 2006.
Wolfe, Cary. "Ecologizing Biopolitics, or What is the 'Bio-' of Biopolitics and Bioart?" In *General Ecology: The New Ecological Paradigm*, edited by Erich Hörl with James Burton, 217–34. London: Bloomsbury, 2017.
Wolfe, Cary. *Before the Law: Humans and Other Animals in a Biopolitical Frame*. Chicago: University of Chicago Press, 2013.

Wolfe, Cary. *What Is Posthumanism?* Minneapolis: University of Minnesota Press, 2010.
Wolfe, Cary. "Human, All Too Human: 'Animal Studies' and the Humanities." *PMLA* 124, no. 2 (2009): 564–75.
Wolfendale, Peter. *Object-Oriented Philosophy: The Noumenon's New Clothes.* Falmouth, UK: Urbanomic, 2014.
Woodward, William A. "Hermann Lotze's Critique of Johannes Müller's Doctrine of Specific Sense Energies." *Medical History* 19, no. 2 (1975): 147–57.
World Health Organization. "Addictive Behaviours: Gaming Disorder." October 22, 2022. who.int/news-room/questions-and-answers/item/addictive-behaviours-gaming-disorder.
Wrangham, Richard. *Catching Fire: How Cooking Made Us Human.* New York: Basic, 2009.
Wundt, Wilhelm. *Grundzüge der Physiologischen Psychologie.* Vol. 1. 12th ed. Leipzig: Engelmann, 1880.
Wundt, Wilhelm. "Philosophy in Germany." *Mind* 2, no. 8 (1877): 493–518.
Wundt, Wilhelm. *Vorlesungen über die Menschen- und Thierseele.* Vol. 1. Leipzig: Leopold Voß, 1863.

Index

Abel, Günter, 91, 96, 191n19, 204n33, 206n52
Acampora, Christa Davis, 155, 213n11
actor-network theory (ANT), xiii, 22, 24, 59, 60, 64
Adorno, Theodor: Heidegger and, 17, 188n35
aesthetic theories, 69, 135, 136
Agamben, Giorgio, 158
Agar, Nicholas, 188n43
agency, xi, xiv, xx, 2, 125, 146, 156, 178; as epiphenomenon, 137; notions of, 133, 134, 135; posthumanism and, 135, 214n25; power and, 219n37; presupposition of, 135; problem of, 214n25
agonistic, xxii, 124, 155, 156, 159, 160, 167, 170, 171, 174, 175, 178
algorithms, 172–73, 220n45
Alphabet Inc., 173
Althusser, Louis, 209n21
animal rationale, xv, 17, 19
animals, 21, 144; calculating, 142–46; discrimination against, 12; eusocial, 58, 59, 64; humans and, 20, 82–83, 112–13
Ansell-Pearson, Keith, 155
ANT. *See* actor-network theory
"Ant-Colony as an Organism, The" (Wheeler), 62

Anthropocene, 26, 57
anthropocentrism, xvi, 6, 14, 65, 77, 99, 125, 208n11; critique of, 166; humanism and, 2; questioning, 64
anthropology, 41; philosophical, 3, 12, 79, 102, 104, 112, 126–27
anthropomorphism, 2, 47, 67
Antichrist, The (Nietzsche), 125, 129
antifoundationalism, 51, 147
antihumanism, 7, 14, 219n34; humanism and, 188n46
Anti-Nietzsche (Bull), 152
anti-Semitism, 17, 200n47, 217n5
ants, 63, 74, 76, 82, 83, 92; behavior of, 70, 86; rhizome of, 200n49; will of, 87
Appel, Frederick, 217n9
Aquinas, 153; truth and, 42, 43, 66
Arendt, Hannah, 156
Aristotle, xvii, 63, 197n16, 197n19, 209n17
Austin, J. L., 138, 220n40
authenticity, 17, 175, 188n35, 221n50
autopoiesis, 50, 128, 194n46, 215n29

Babich, Babette, 206n2, 216n38
Badmington, Neil, 7, 101, 188n44
Baecker, Dirk, 22, 116, 175, 176, 211n48
Balsamo, Anne, 188n44

241

Barad, Karen, 54, 128
barbarism, xix, 9, 16; media-technological, 114–21
Bateson, Gregory, 22, 79, 118
Bau und Leben des socialen Körpers/ Formation and life of the social body (Schäffle), 71
Baudelaire, Charles, 95
Baudrillard, Jean, 220n40
Beard, Charles A., 107, 210n25
Beaufret, Jean, 16
beehive, 63, 70, 71, 82, 85, 202n17; language and, 81; swarming from, 72–77
Before the Law (Wolfe), 4
behavior, 94, 176; animal, 88; communicative, 179; insect, xvii, 70, 86; modifications, 141; patterns, 82, 173; social, 8, 61, 70, 86, 91, 203n23; swarm, 8, 58, 169, 173
Beiner, Ronald, xxi, 152, 154, 155, 218n15
Being, truth of, 188n40
Being and Time (Heidegger), 188n35
Benjamin, Walter, 220n40
Bennett, Jane, xi, xviii, 19, 44, 54, 80, 94; materialism of, 200n1, 205n41
Bergson, Henri, 196n64, 205n41
Bertino, Christian, 82, 198n26, 211n42
Beyond Good and Evil (Nietzsche), 39, 47, 48, 59, 90–93, 151, 211n47, 218n23; free spirit and, 110; willpower and, 90
binaries, onto-theological, 19, 80, 94
biological systems, 70, 71, 176–77, 181
biopolitics, 128, 130, 147; affirmative, xviii, 64, 158, 167, 181, 187n12; beyond, 157–60; humanism and, 15
Bios (Esposito), xxi, 4, 157
biosphere, 57, 112
biotechnology, xix, 3, 22, 188n42, 188n43
Birth of Tragedy, The (Nietzsche), 49
Blade Runner, 123
Bloomsbury Handbook of Posthumanism (Thomsen and Wamberg), 11
Boltanski, Luc, 181
Bonabeau, Eric, 60, 172
bonds: cultural, 10; normative/emotional, 179; social, 10
Bonnet, Charles, 62
Bostrom, Nick, 53, 101, 188n43; on Nietzsche, 186n8; transhumanism and, 2, 14, 52
Braidotti, Rosi, xxii, 6, 7, 11, 15, 26, 126, 167, 177, 187n21, 188n44, 189n46, 219n34, 219–20n38; material feminism and, 61; posthumanism and, 128, 146, 166, 188n46; technology and, xxi, 104, 147, 148; togetherness and, 159, 168; on tradition/customs, 127
Brobjer, Thomas H., 185n4
Brun, Rudolph, 62
Buchanan, Allen, 188n43
Büchner, Ludwig, 192n24
Bugnion, Edouard, 62
Bull, Malcolm, xxi, 152
Burckhardt, Jakob, 114
Butler, Judith, 25, 130, 149, 159

calculus: economic, xxi, 131, 142, 143, 144, 145, 149, 163, 167;

INDEX 243

human, 145; psycho-economic, xx, xxii, 142, 144, 150, 161, 162, 163, 180, 219n30
Callon, Michel, 22
Cambridge Companion to Literature and the Posthuman (Clarke and Rossini), 11
Capital (Marx), 106, 209n22
capitalism, 163, 164, 210n25; education/science and, 120; values, 164
Carbonell, Curtis, 11
cartography, 11, 22, 23, 24, 52, 166, 215n29
Case of Wagner, The (Nietzsche), 205n45
Caspari, Otto, 87
causality, 37, 38, 170, 190n8; attribution of, 45, 46; concept of, 48; considerations of, 194n41; primacy of, 33, 45
changeling (*Wechselbalg*), 133
Chladni, Ernst, 34, 35, 36, 38, 192n25
Christianity, 15, 17, 125, 214n19; humanism of, xv
civilization, 207n6; in counterdistinction, 19; history of, 194n42; process, 143, 161; technology and, 147
Civilization and its Discontent (Freud), 59
Clark, Maudemarie, 91, 191n22
Clarke, Bruce, 11, 57, 58
climate change, xxii, 159, 163, 164
cognition, xiv, xvi, xvii, 9, 38, 42–51, 60, 89, 118; capabilities/limits of, 43, 46; faculty of (*Erkenntnisvermögen*), 39; neocybernetic theories of, 36; physiology of, 30–36
cohesion, 71, 72, 76, 115, 135
Colebrook, Claire, 11, 18, 49
collectives, xvii–xviii, 63–64, 86, 87, 98–99, 198n29; animal, 75; human, 62, 75, 83; formation of, 72, 74, 76
communication, 56, 83, 98, 119, 136, 141, 183; abilities, xvii, 60; patterns, 62, 121, 173; process of, 87, 179; social medium of, 172; technology of, 136
community, 182; posthuman, 159, 165–76; sense of, 174; term, xxii, 160, 169
competition, 92, 174; affirmation of, 167
Comte, Auguste, 70
Conflict and Contrast in Nietzsche's Philosophy (Siemens and Pearson), 151
Connolly, William E., 155, 213n11
conscience, genealogy of, 137–42, 215n34
consciousness, 31, 43, 66, 90, 109, 129, 136, 178; computer-generated, 52; higher, 54; language and, 38, 41–42, 50; temporal structure of, 113, 114; transcendental, 44, 128, 130, 136; understanding of, 86
constructivism, 147; operational, 36, 56, 57
correlationism, xvi, 30; embodiment and, 51–58; Kantian, 24, 27, 55; materialism and, 26
cosmopolitanism, 19, 49

Cowan, Michael, xix, 88, 95, 203n30
Crary, Jonathan, 29–30, 191n22
Crawford, Claudia, 82, 193n36
crimes, 74, 142, 163, 211n46; against humanity, 175
Critical Posthumanism (Herbrechter), 65
cruelty, 141, 144, 163; intellectualization and "deification" of, 140
cultivation, xx, 104, 111, 121, 144, 150, 162, 165; human, 164; pain and, 141; technology of, 124, 127; theory, 140
Cult of the Will (Cowan), 94
cultural decline, 79, 114, 206n52, 211n48
cultural production, 8, 10
cultural studies, 103, 105, 208n11
culture: agricultural heritage of, 104; book, 121; changes in, 115; cults of knowledge and, 211–12n49; definition of, 211n45; Greek, 84, 114, 115, 116; human nature and, 115; life/thought/appearance/will and, 115; mass, 163; modern, 114, 117; nature and, 116, 127, 148; newspapers and, 211n43; oral, 173; postbook, 10; posthumanism and, 116; premodern, 212n49; Roman, 116; technology and, 111; Western, 123–24
cybernetics, 7, 27, 61, 62, 196n3, 197n12
"Cyborg Manifesto" (Haraway), 123, 124, 207n4
cyborgs, xix, xx, 7, 101–2; *Übermenschen* and, 121–24

Damasio, Antonio, 192n26
Dangerous Minds: Nietzsche, Heidegger, and the Return of the Far Right (Beiner), 152
Darwin, Charles, 63, 202n20; instinct and, 84, 85; law of natural selection and, 84; social realm and, 170
data mining, 220n45; algorithmic, 172–73
Daybreak (Nietzsche), 25, 73, 74, 98
Death of the Post-Human (Colebrook), 11
décadence, 95, 97, 205n45
deconstruction, xii, xviii, 1, 13, 21, 47, 94, 136, 147, 208n12; postmodernist, 26
Deleuze, Gilles, xvii, xxiii, 23, 44, 76, 80, 196n64, 200n49, 205n41, 219n34, 220n40; desire and, 84; material vitalism of, xviii; posthumanism and, 61
democracy, 151, 152, 156, 157; conception of, 155; revitalizing, 155
Derrida, Jacques, xxiii, 15, 64, 98, 105, 215n29, 220n40; on Nietzsche, xix, 185n3; observation and, 194n47; posthumanist theory and, 136; spider analogy of, 197n21
Der thierische Wille/The animal will (Schneider), 87, 202n17
de Saussure, Henri, 62
Descartes, René, 194n43
Des sociétés animals/Die thierischen Gesellschaften/On Animal Societies (Espinas), xvii, 62, 70–72, 73, 87
Detwiler, Bruce, 217n9
development: conceptual, 124; language, 82, 83, 86; social, 170;

socio-economic, 27; technological, 112
dichotomies, xix, 39, 44, 45, 64, 69, 80, 85, 93, 94, 95, 142, 148, 164, 185n8, 190n14, 199n32, 218n23; dominant, 156; nature–culture, 116, 127; political, 157; reinforcing, xi; social theories and, 98
Diderot, Denis, 205n41
Die Gesellschaft der Gesellschaft/The society of society (Luhmann), 221n59
Die Lehre von der Tonempfindung/*On the Sensation of Tone as a Physiological Basis for the Theory of Music* (Helmholtz), 31, 192n25
Die Sprache als Kunst (Gerber), 81
Die Wissenschaft der Gesellschaft (Luhmann), 50
differentia specifica, humans/animals and, 82–83
differentiations, 36, 56, 57, 58, 165, 177, 179; functional, 51; social, 51, 80, 178; technological, 178
discipline, 178; formation of, 11; physiological, 48; as technology, 140–41
Discipline and Punish (Foucault), 140
Discourse Networks 1800/1900 (Kittler), 111
discrimination, xv, 12; political modes of, 156
disgregation, 97, 98, 99, 205n45, 206n52
"'Disgregation des Willens'— Nietzsche über Individuum und Individualität" / "'Disgregation of the Will'— Nietzsche on the Individual and Individuality" (Hamacher), 96
diversity, 3, 128, 129, 180
Dombowsky, Don, 217n9
domination, xxii, 72, 153, 174; modes of, 166; political, 64, 97
Dorigo, Marco, 60, 172
Dries, Manuel, 130
Driesch, Hans, 205n41
drive, 67, 79, 88; energy/gravity for, 93; fundamental, 69; unconscious, 204n32
Drochon, Hugo, 155, 156
dualism, xiii, 99; mind-body, 2, 24, 53, 54; Platonic, 121, 123
du Bois-Reymond, Emil, 28
Dudrick, David, 91
Dugin, Aleksandr, 152
Dühring, Eugen, 162
Dunkle, Ian, 204n40
Durkheim, Émile, 199n31

Ecce Homo (Nietzsche), 1, 152
ecological, 177; battles, xvi, xvii, 56, 57, 165; constraints, 158; modes of thinking, 180; paradigm, 159; position/vantage point, 12, 85; self, 190n14
ecological modes, 159, 180
economic system, xxii, 74, 159, 164
education, 73; capitalism and, 120; mass, 73, 115
educational operations, 115, 121; labor market and, 119–20; modern, 149
egalitarianism, 156, 174, 219n28
Elements of a Philosophy of Technology (Kapp), 110, 136

elitism, 17, 165, 174
embeddedness, 52, 127–31, 146, 172, 175, 179; biological, 128; cognitive, 160; cultural, 120; environmental, 65; material, 95; social, 2; technological, 120, 128, 130, 164, 177
embodiment, 127, 128; correlationism and, 51–58; enactive, 30, 35, 51, 170; posthumanism and, 51; radical, 30, 51; social, 72, 86
Emden, Christian, 27, 28, 144, 145, 191n17, 192n24, 212n4; on naturalism, 126; physiology and, 34
Enlightenment, xv, 5, 13, 14, 15, 18, 19, 20, 49, 66, 73, 114, 117, 121, 156, 168, 175; cognition and, 31; dogma of, 119; Nietzschean philosophy and, 152
entomology, xii, xvii, 58, 72, 85, 93, 98, 126, 169, 175; conceptual challenges of, 62–65; human collectives and, 62; posthumanism and, 61–62, 72, 76
environment, 84, 169; material, 60; political, 181; social, 146, 148, 181; technological, 146, 148, 181
environmental challenges, 8, 26, 57, 58
environmental changes, 76, 220n43
Epicuris, 205n41
epistêmê, xvi, 23, 30, 36, 51, 105; *technê* and, 208–9n17
epistemology, xiv, 2, 6, 21, 32, 36, 126, 153, 165, 191n19, 193n39, 196n64; challenges for, 50; constructivist, 24, 55, 57, 153; critical, 12; cybernetically informed, xvi; Kantian, 27; neocybernetic, xvii, 57; neo-Kantian, xvi, 26, 27; posthumanist, 42, 128; poststructuralist, 26; post-truth, 42–51

Escherich, Karl, 76
Espinas, Alfred, xvii, 62, 63, 73, 87, 198n29, 199n31, 199n32, 201n13, 202n16, 202n17; Huber and, 72; morality and, 199n29; Nietzsche and, 70–72; social embodiment and, 72
Esposito, Roberto, xxi, 159, 168; Nietzsche and, 2, 4, 157, 158
essence, 20, 21, 61, 73, 81, 141, 188n34, 188n40, 209n21; of humanism, 17, 19; of instinct, 79; rhetorical, 13; of things, 31, 35, 47
ethics, xxi, 5, 129, 166; environmental, 159; hyperhumanist, 168; posthuman, 128, 150, 160, 165, 176–83; utilitarian, 164
"Ethics for Cognitive Assemblages" (Hayles), 180
ethological research, xii, xiii, 58, 70, 76
ethos, 1, 14, 69, 152, 157, 215n29; anti-authoritarian, 166; authoritarian, 61; philosophical, 14; political, 3; posthumanist, xiv, xv, xxii, xxiii, 6, 16, 144, 160, 165, 182, 219n30
Euripedes, 16
Evola, Julius, 152
evolutionary theory, xiii, 86, 112
Ex Machina, 123

fascism, 153, 170, 176; Nietzsche and, 152, 154–55, 156, 217n6

Fechner, Gustav, 54, 189n4, 192n24
feedback loops, 169, 176; positive/negative, 87, 172, 190n14
feminism, 61, 128, 152; socialist-, 124
Ferrando, Francesca, 44, 128, 130, 166, 193n39
Fichte, Johann Gottlieb, 24
Foerster, Bernhard, 75, 200n47
Forel, Auguste, 62, 70, 76, 197n12
Fornari, Maria Cristina, 70, 198n29
Foucault, Michel, xv, 14, 15, 23, 105, 110, 149, 156, 220n40; biopolitical perspective of, 128, 130; cultivation theory and, 140; on discipline/technology, 140–41; Nietzsche and, 157; posthumanism and, 157; power *dispositifs* and, 220n42; work of, 25
Fouillee, Alfred, 198–99n29
foundational fact (*fundamentale Thatsache*), 89
freedom, 73, 81, 89, 97, 132, 133, 198n27, 214n25, 218n15; awareness of, 90; individual, xxi, 153
Freud, Sigmund, 59
Fuchs, Thomas, 35, 170, 190n14, 192n26
Fukuyama, Francis, 14, 101
Funke, Otto, 192n24

Gadamer, Hans-Georg, 220n40
Gaian Systems (Clarke), 57
Gaian thought, 57, 58
Gay Science, The (Nietzsche), 97, 162, 173
gender, xx, 207n4; concept of, 25; norms, 123; perceptions of, 25; questions of, 123; studies, 148
"Genealogy of Posthumanism, A" (Herbrechter), 2
genius, 58, 107, 140; aesthetics of, 134; defining, 134–35
Gerber, Gustav, 81
Geschichte des Materialismus/History of Materialism (Lange), 28, 43, 192n24, 197n19, 199–200n44
Goethe, Johann Wolfgang von, 16, 58, 115, 190n8, 221n50
Gordon, Deborah, 61, 62, 196n3
Graham, Elaine L., 7, 21, 101, 122, 188n44, 212n55
grammar, 96, 134, 149
Gray, Chris Hables, 188n44
Grosz, Elizabeth, 54
Grundzüge der physiologischen Psychologie/Principles of Physiological Psychology (Wundt), 88, 89
Guattari, Félix, xvii, xxiii, 23, 44, 76, 80, 200n49, 205n41, 219n34; desire and, 84; material vitalism of, xviii; posthumanism and, 61
guilt (*Schuld*), 142, 180
Guyau, Jean-Marie, 198n29

Habermas, Jürgen, 152, 217n8
Haeckel, Ernst, 28, 189n4
Hamacher, Werner, 96–97, 97–98, 176, 206n54
Handbuch der Physiologie des Menschen für Vorlesungen/Elements of Physiology (Müller), 28, 31
Haraway, Donna, 7, 101, 123, 128, 188n44, 208n12; autopoiesis and, 194n46; cyborg figure of, xx, 124, 207n4; posthumanism and, 3; socialist-feminism and, 124; study by, 9

INDEX

hard sciences, 10; knowledge from, 44; language and, 133
Harman, Graham, 7
Harris, John, 188n43
Hartmann, Eduard von, 63, 71, 81, 82, 192n24, 201n3, 201n5, 201n7; instinct and, 83, 85; language and, 84
Hatab, Lawrence, 145, 146, 155, 213n11; adversarial system and, 156; sovereign individual and, 216n35
Hauskeller, Michael, 11
Havelock, Eric A., xx, 110, 121
Hayles, N. Katherine, xv, 5, 7, 14, 101, 131, 173, 188n44, 195n57, 208n12; on cognitive nonconscious, xvii, 24, 54, 55; embodiment and, 51, 52; on human autonomy, 180; posthumanism and, 3
Hegel, George Wilhelm Friedrich, 195n64, 221n50
Heidegger, Martin, xv, 5, 16–17, 21, 53, 105, 122, 207n6; Enlightenment and, 19; humanism and, xv, 17, 19; jargon of authenticity and, 188n35; Nietzsche and, 18, 187n34; observation and, 194n47; ontology of, 188n34; political stance of, 17–18
Heidelberger, Michael, 190–91n16, 194n43
Heit, Helmut, 212n4
Helmholtz, Herman von, 27, 28, 29, 30, 54, 189n4, 191n16, 191n17, 192n24, 192n25; Nietzsche and, 203n23; on sensations, 31
Henle, Friedrich Gustav Jakob, 28
Herbrechter, Stefan, xv, 3, 7, 9, 10, 11, 14, 65, 121; on Nietzsche, 2, 186n7; posthumanism and, 1, 13, 22, 52–53; transhumanism and, 122
herd, 63, 64, 66, 67, 72, 80, 94, 85, 86, 165, 173, 176; dissolution of, 97; formation of, xix, 87; individual and, 199n32; will and, 94–98
Herder, Johann Gottfried, 82
"History of Transhumanist Thought" (Bostrom), 2
"Hitler Swarm, The" (Baecker), 175
Hlavajova, Maria, 6, 11
Hobbes, Thomas, 138, 197n19, 205n41, 215n33
Holbach, Paul Heinrich Dietrich, 27
Hölderlin, Friedrich, 16
Holub, Robert C., 70, 217n5; Nietzsche and, 185n4, 218n21; organicist sociology and, 199n30; social question and, 73, 200n45
hominization, xix, xx, 104, 111, 118, 121, 122, 132, 165; process of, 109, 114, 119; technology of, xiv, 124, 127, 131, 141–42, 149
homo, 17
homo animalis, 20
homo barbarus, 16
homo humanus, 16
homo natura, 148, 167
Homo Natura (Lemm), xxi, 104, 146
Hörl, Erich, 159
How We Became Posthuman (Hayles), 5, 13–14, 51, 54
Huber, François, 62
Huber, Pierre, 62, 72, 84

Huddleston, Andrew, 214n19
Hughes, James, 188n43
Human, All Too Human (Nietzsche), 4, 38, 45, 73, 218n23
human dignity, 153, 182
humaneness, 17, 74, 129, 143, 154, 161, 180
humanism, xv, 9, 41, 117, 121, 129, 150, 163, 164, 175, 177; analysis of, 17; anthropocentrism and, 2, 65, 77; antihumanism and, 188n46; coherence of, 19; core of, 80; described, 13–21; Enlightenment, 168; Other and, 19; past/present, 14; posthumanism and, xi, 3, 15, 21; profiling, 16; rejection of, 18; secular liberal, 122; tendencies of, 15
humanitarian crises, 155, 175, 182
humanity, 7, 14, 74, 129, 149, 165
human nature, xxiii, 130, 149, 211n46; culture and, 115; essentialization of, 126; notion of, 22; reconstruction of, 148
Human Nature in an Age of Biotechnology (Sharon), 21–22, 53
humans: autonomous, 1, 8, 12, 13, 52, 56, 66, 135, 139, 180; animals and, 20, 80, 82–83, 112–13; cultivation of, 103; decentering of, 8; embeddedness of, 164; humaneness of, 17; knowledge and, 65; posthumanist, 139; self-assessment of, 83; status of, xiv; technical and, 109
human sociality, xii, xviii, 58, 72, 76
Humboldt, Alexander von, 115
Humboldt, Wilhelm von, 81, 201n3
Hume, David, 137, 161

I, synthetic concept of, 90
idealism, 221n50; ethical, 168; neo-Platonic, 19–20, 21; Platonic, 19–20, 21, 122
identity, 146; assumptions of, 46; durability of, 146; false sense of, 41; language and, 46
Ihde, Don, 22
imagination, 10, 31, 37, 123, 214n25; fiery liquid of, 68; language and, 68; moralistic, 160; plastic, 58
immunitary *dispositifs*, 157, 158, 159
immunization, 158, 159
individual: destruction of, 96; material environment and, 60; normalization of, 97; society and, 95. *See also* sovereign individual
individualism, 64, 77, 176
individuality, 94, 173; affirmations of, 96–97; common, 98
inequality, xxii, 145, 159, 163
information: flow of, 119; processing, 118, 119; theory, 62, 118; unchecked, 117
insects, 63, 72–73, 75, 80, 83, 92; behavior of, xvii, 70; colony-forming, 60; reference to, 79; sociality of, xvii, 70–72; writings on, 65
instinct, 61, 67, 79; automatism of, 84; biological/social, 83; concept of, 80, 83; herd, 84; language and, 67, 83; malleable, 80–85; posthumanist notion of, 83; understanding of, 80
interactive processes, 70, 72, 86

Judeo-Christian tradition, 16, 19, 98

Kant, Immanuel, xvi, 14, 24, 85, 194n43, 201n7; aesthetics of genius and, 134; consciousness and, 136; correlationism of, 26; Enlightenment and, 5; epistemology of, 26, 27; Hume and, 137; instinct and, 79; posthumanism and, 13

Kantian catastrophe, 2, 55

Kapp, Ernst, xix, 109, 110, 136

Katsafanas, Paul, 213n13

Kirkwood, Jeffrey, 108

Kittler, Friedrich, 105, 131, 189n4, 207n6, 208n12, 211n47; loss of meaning and, 115; media theory and, 121; typewriter and, 111

knowledge, 117; cults of, 211–12n49; distribution, 115; historical, 116; humans and, 65; limits of, 44

Krämer, Sybille, 103

Kroker, Arthur, 207n6

Kulturtechniken, 103, 104, 208n12

Kulturtechnologie, 103

Kunnas, Tarmo, 207n6

Kurzweil, Ray, 14, 101, 123, 188n43, 212n56

Kutter, Heinrich, 62

labor market, educational operations and, 119–20

Lambrecht, Georg Friedrich von, 105, 209n18

Lang, Fritz, 209n22

Lange, Friedrich Albert, 28, 43, 63, 192n24, 193n36, 200n44; atomism and, 47; on Hobbes, 197n19

language, 34, 36, 55, 99; beehive and, 81; conceptual world of, 43; consciousness and, 38, 41–42, 50; creative force of, 133–34; deeds and, 133; development of, 82, 83, 86; genesis of, 35, 191n21; hard sciences and, 133; humanist conceit about, 67; identity/matter and, 46; imagination and, 68; instinct and, 67, 83; mental image and, 33; metaphorical, 69; neurological responses to, 70; neurophysiological effects of, 65–70; organicist model of, 37–42; perception and, 68; physiology and, 39; posthumanist understanding of, 127; power of, 149; as putative science, 40; seductions of, 38, 133; social formations of, 65–70; structure of, 46; as technology, 131–37, 214n19

Latour, Bruno, xix, 11–12, 22, 23, 64, 76, 154, 178, 215n29; ANT and, xxiii, 24, 59; morality and, 179–80; posthumanism and, 177; technology and, 216n36

Law, John, 22

law of natural selection, 84

Lectures on Latin Grammar (Nietzsche), 79, 81

Leiber, Theodor, 191n16

Leibniz, Gottfried Wilhelm, 204n33

Leitner, Brian, 217n5

Lemm, Vanessa, xxi, 44, 64, 146, 148, 159, 166, 168, 193n37, 207n5, 219n34, 219n37; biopolitics and, xviii, 167; naturalism and, 213n4; Nietzsche/posthumanism and, 97, 103–4, 147, 187n12, 206n53; shared debt and, xxii; technology and, 104

Leroi-Gourhan, André, 98
Les Fourmis de la Suisse (Forel), 62
Letter on Humanism (Heidegger), xv, 16, 17, 18, 19, 20, 21, 188n35, 188n40
Leviathan (Hobbes), 91
Lewes, George Henry, 71–72
liberalism, 152, 153, 156, 163, 217n4
life, 174; conception of, 92; decoration of, 114; matter and, 80; nature of, 92
life sciences, xiii, 38, 112, 130
Lipps, Theodor, 192n27
literacy rates, 115, 141, 173
Little Red Riding Hood, 116–17
Locke, John, 36, 138, 156
logic, 20, 39, 41, 49, 50, 82, 158, 167, 168, 206n55; bivalent, 191; circular, 201n13, 204n32; compensatory, xxi, 145, 163; humanist, 174; situatedness of, 130
Lotze, Rudolf Hermann, 27, 30, 192n24
Lucretius, 205n41
Luhmann, Niklas, xxiii, 2, 15, 22, 50, 51, 215n29, 221n59; constructivism of, 195n47; embeddedness and, 179; morality and, 179–80; Nietzsche and, 194n47; posthumanism and, 136, 177; systems theory and, 178, 220n42; technology and, 164
Lyotard, Jean-François, 217n6, 220n40

manas, 142–46
Mängelwesen, 147, 207n5
Margulis, Lynn, 128

Marx, Karl, xix, 17, 110, 164, 200n44, 207n6, 209n21, 209n22, 210n25; analysis by, 119; capitalism and, 120; machinery and, 120; Nietzsche and, 124
Marx, Leo, 206n3
Marxism, xv, 15
mass media, xix, 9, 114–17, 173; destructive effects of, 121; information processing and, 118; posthuman and, 10
Massumi, Brian, 61, 83–84
materialism, xii, xxiii, 7, 53, 54, 56, 57, 131; correlationalism and, 26; ontological contentions of, xvi; posthumanism and, 24, 52, 53
Maturana, Humberto, xvi
McLuhan, Marshall, xx, 110, 121
Meaning of Life, The (Uexküll), 190n8
media: communication, 8, 9; coverage, 119; electronic/digital, 10, 103, 121; histrionics of, 111; social, 119, 173, 193n32; studies, 103; theory, 110, 121. *See also* mass media
Meillassoux, Quentin, 7, 53, 195n58, 195n60; anti-rationalist doctrines of, 196n64; on correlation, 55; Kantian catastrophe and, 2, 55; on subjectivity, 195n64
Mellamphy, Dan, 197n20
Mellamphy, Nandita Biswas, 197n20
memory, 118, 141; humans/animals and, 112–13; increase in, 141
mental images, 34, 37, 38, 47; language and, 33; nerve stimulus and, 33

mercy, 143, 144, 145, 161–62
Meredith, Thomas, 215n34
Meta, 173
"Metalogue: What Is an Instinct?" (Bateson), 79
Metamorphosis of Plants (Goethe), 190n8
metaphor, 32, 35, 41, 67, 114; concept of, 47; language and, 69; truth and, 68
metaphysics, 15, 45, 187n34, 195n64, 204n40; neo-Platonic, 20, 122; Platonic, 18
Metropolis (Lang), 209n22
migration, xxii, 159, 163, 182
Mind (journal), 202n20
Miyasaki, Donovan, 156, 174, 219n28
Modernism, 14
modernity, 98; culture and, 114; *décadence* of, 95; Western, 212n49
Moleschott, Jakob, 27
Moore, Gregory, xiii, 111
moral codes, 2, 164, 168, 175, 176
morality, xiv, 39, 92, 173, 205n49, 220n46; critique of, xxii, 145, 160, 180, 182, 194n42; history of, 169; Judeo-Christian, 3; law and, 145; recognizing, 179–80; reduction of, 181; slave, 149, 203n29
morality of customs (*Sittlichkeit der Sitte*), 97
morals: genealogy of, 142, 167; Judeo-Christian, 153
moral values, 129, 131
Moravec, Hans, 14, 53, 188n43
More, Max, 101, 123
Morton, Timothy, 7, 26, 56

Mouffe, Chantal, 218n18
Müller, Jan-Werner, 156, 160
Müller, Johannes Peter, xvi, 27, 32, 38, 48, 50, 58, 191n17, 194n43; fantasy and, 192n28; Kantian catastrophe and, 55; lecture by, 190n8; nerve energies and, 34; neurophysiology and, 45; Nietzsche and, xvi, 37, 57; optics and, 192–93n31; physiology and, 30–31, 193n36; research by, 55–56, 189n5; theorem of, 29–30; work of, 28–29, 33, 35, 36, 37
myrmecology, 59, 61–62

nationalism, 152, 200n47; countering, 162–63
National Socialism, xv, 15, 156
Natural History of Ants, The (Huber), 72
naturalism, 126, 147, 213n4
natural sciences, 16, 58, 126, 185n4, 189n3
nature, 115, 134, 150, 195n64; culture and, 116, 127, 148; generative/creative force and, 148; posthumanism and, 116, 148; technology and, 147, 148
Nayar, Pramod, 7, 8, 13
Nazism, 3, 17, 157, 158, 175, 188n35
neocybernetics, xxiii, 24, 37, 50, 53, 197n12
neo-Platonism, 16
neopragmatism, 177
neovitalism, 158
nerve stimulus, 30, 32–33, 34, 35, 38, 47–48, 55
nervous system, 33, 34, 45, 68

networks, 22, 34, 60, 95, 175, 178, 196n3; discourse, 111; dynamics of, 178; material, xxiii, 99, 165; neural, xvi, 30, 37, 55; social, xxiii, 99, 165; technological, 99, 165
Neumann, Constantin Georg, 88
neurology, 37, 68, 70, 80
neurophysiology, xii, 9, 27, 32, 34, 38, 45, 55–56, 126
neuropsychology, 67, 68, 69, 86, 154
Nietzsche, Friedrich: concept of the organic since Kant, 27, 31; conservatism of, 10; philosophy of, 3, 5, 6, 23, 63, 70, 152, 154–55, 156, 182, 199n29, 217n8; political legacy of, 5, 162, 187n15; styles of, 185n3, 197n21; writing of, xi–xiii, xiv, xvii, 17, 218n23
Nietzschean Bestiary (Acampora and Acampora), 63
Nietzsche on Language, Consciousness, and the Body (Emden), 27
Nietzsche, Power and Politics (Siemens and Roodt), 151
Nietzsche's Corps/E (Waite), 207n6
Nietzsche's Naturalism (Emden), 212n4
nihilism, 51, 163, 176, 205n45
nonconscious, 63, 79; cognitive, xvii, 24, 54, 55

Object-Oriented Ontology (OOO), 7, 24, 53, 195n60
Ong, Walter J., xx, 110
On the Diversity of Human Language Construction and Its Influence on the Mental Development of the Human Species (Humboldt), 81
On the Genealogy of Morals (Nietzsche), xx, 88, 110, 110–11, 127, 131–33, 136, 137, 139, 140–44, 149, 150, 155, 160–62, 169–71
"On the Uses and Disadvantages of History for Life," 112–20, 150, 211n46, 211n47, 216n35
ontology, xvi, 18, 44, 56, 57, 61, 188n34, 219n30
On Truth and Lies in a Nonmoral Sense (Nietzsche), 31, 42, 47, 48, 55, 65–69, 80, 82, 96, 154, 191n17, 193n33, 193n34, 198n27
OOO. *See* Object-Oriented Ontology
organic activity, 37, 88
Other, 20; barbaric, 19; cognitive/moral authority over, 21
overhuman, 1, 2, 3, 5, 7, 101, 104, 110, 122, 174, 186n8, 210n35, 212n56
Owen, David, 134, 135, 155, 156

pain: avoiding, 87; cultivation and, 141; understanding of, 86
Palgrave Handbook of Critical Posthumanism (Herbrechter), 11
Palgrave Handbook of Posthumanism in Film and Television (Hauskeller, Philbeck, and Carbonell), 11
panhumanism, 7
Parikka, Jussi, xvii, 54, 61, 83
Parisi, Luciana, 54
Parry, Richard, 209n17
Patton, Paul, 129, 152, 217n9
Pearson, James, 151
penal codes, 139, 143, 162, 169, 170
Pepperell, Robert, 1

perception, 88; assumptions and, 57–58; language and, 68
perspectivism, 126, 128, 195n47, 213n12; affirmation of, 129; monistic, 93
phenomenology, 32, 34; of posthuman, 8–12
Philbeck, Thomas, 11
Philosophical Discourse of Modernity, The (Habermas), 217n8
Philosophical Posthumanism (Fernando), 128
Philosophie des Unbewussten/Philosophy of the Unconscious (Hartmann), 71, 81
philosophy, xiv, 165; animal, 3; historical, 39; moral, 199n29
Physical Basis of Mind, The (Lewes), 71–72
physiology, xii, 8, 20, 27, 29, 33, 34, 37, 44, 49, 50, 51, 55, 64, 87, 90, 93, 126, 137; experimental, 112; human, xvi, 102, 208n8; idealist, 47; language and, 39; of sight, 193n36; study of, 47–48
Pickering, Andrew, 7, 22
Pippin, Robert, 2, 134, 135
Plato, 16, 17, 36, 209n17
Platonic theory, 17, 18, 19, 21, 122
Plato's Doctrine of Truth (Heidegger), 16–17, 21
Plessner, Helmuth, 104, 109, 146, 207n5
pluralism, 3, 127, 128, 129, 146, 182
political change, 56, 175
political conflict, 73, 166
political thought, 4, 76, 151, 155, 204n35

politics, 161, 179; aestheticization of, 147; mental processes and, 91; moralistic imagination of, 160; right-wing, 154, 157
"Politische Steuerungsfähigkeit eines Gemeinwesens" / Political ability to steer a commonwealth (Luhmann), 179
populism, xxii, 156–57, 160, 161
posthuman, xxi, 177; cultural/media representations of, 10; ethical/political contentions of, 12; phenomenology of, 8–12; posthumanism and, 8
Posthuman, The (Braidotti), 15, 188n46
Posthuman Glossary (Braidotti and Hlavajova), 11
posthumanism: assemblage, 7, 146, 147, 166; cartography of, 22, 125; contemporary, xii, xiv, xxii, 5, 24, 44, 128, 129, 157; critical, xv, xxi, xxiii, 5, 7, 13, 20, 21, 22, 27, 52, 53, 56, 60, 125, 127, 156, 180, 182, 183; described, 6–8, 11, 188n46; epistemological/social/technological aspects of, 165; ethnopolitical dimension of, 4; great divide of, 21–24; inquiry into, xxi, 12, 13, 153; introductions to, 9, 52–53; literature on, xiv, 11, 61; mediated, 7, 188n42; methodological, xxiii, 7, 22, 23, 27, 50, 51, 61, 160, 166, 176–77; Nietzsche and, xiv–xv, 1–2, 3, 4–5, 20; philosophy of, 6, 7, 128, 130; politics of, 157, 160, 187n15; radical, 7, 22, 23, 50; reading, 5–6, 8; social/political

INDEX 255

relevance for, 124, 165; strands of, xv, xxiii, 7, 26, 98, 122–23, 125, 126, 182, 216n40; technological aspects of, 165
Posthumanism (Herbrechter), 1, 52–53
"Posthumanism and Community Life" (Lemm), 104
Posthumanism Glossary (Braidotti and Hlavajova), 6
"Posthumanism Manifesto" (Pepperell), 1
posthumanist thought, xvii, 7, 21, 58, 127
postmodernism, 7, 14, 44, 52, 170, 220n40
poststructuralism, xxiii, 7, 21, 24, 27, 52, 64, 83, 131, 156, 177; Nietzsche and, 25, 26; posthumanism and, 50
power, 92; affirmations of, xxii, 165–66, 171; agency and, 219n37; consolidation of, 161, 162; political, 74; redistribution/refinement of, 140; truth and, 25; will to, 85, 93. *See also* willpower; will to power
presuppositions, 12, 18, 41, 89, 97, 125, 130, 135, 137, 157; anthropological, xvii, 63; biological, 92; metaphysical, xiii, 17; naturalist, xiv; poststructuralist, 147; religious, 66
Principles of Sociology (Spencer), 71
printing press, xx, 9, 103, 121, 141, 173
psychology, 8, 37, 64, 87, 91, 93, 94, 134, 162, 178, 204n40; animal, 88; empirical, 9

psychophysiology, 87, 90, 92, 94, 95
punishment, 143, 162, 171; conception of, 144; corporeal, 141; modes of, 170; public, 140
Putin, Vladimir, 152

Rajan, Tilottama, 216n40
rationality, 82, 117; Enlightenment, 49; thermodynamic, 19, 49
Rawls, John, 156, 215n33
reality: external, 42; physiological, 41, 57; postmaterial, 53; psychological, 57
Reassembling the Social (Latour), 59, 178
"Recent Work on Nietzsche's Social and Political Philosophy" (Patton), 152
recursion, 49, 50, 148
reductionism, 165; ontological/materialist, xvii
Rée, Paul, 202n20
relationality, xxii, 61, 99, 153, 165, 166
relationships, 90; causal, 170; creditor–debtor relationships, 161; political, 161
relativism, 26, 42, 153
Renaissance, 14, 19
resentment, xx, xxii, 99, 132, 165, 178, 219n30; combatting, 145; politics of calculation and, 160–64; sociopolitical, 219n30
respectfulness, 3, 129, 213n11
responsibility, 8, 132, 167, 168, 177, 178
ressentiment, 160, 161
Reuter, Sören, 203n23

INDEX

rhetoric, 69, 117, 155; activist, 177; post-truth, 26
Richardson, John, 92, 93, 136, 137, 173, 205n49, 213n12, 220n46
Roden, David, 52
Rolph, W. H., 70
Roodt, Vasti, 151, 155
Rorty, Richard, 177
Rossini, Manuela, 11
Rousseau, Jean-Jacques, 138
Roux, Wilhelm, 70, 87, 199n30

Santschi, Félix, 62
Savulescu, Julian, 188n43
Schäffle, Albert, 71
Schatzberg, Eric, 107, 210–11n25
Schelling, Friedrich Wilhelm Joseph, 195n64, 204n33
Schiller, Friedrich, 16
Schleiden, Matthias Jacob, 27
Schlüpmann, Heide, 211–12n49
Schmitz-Dumont, Otto, 87
Schnädelbach, Herbert, 191n22
Schneider, Georg Heinrich, 63, 70, 87–88
Schopenhauer, Arthur, 38, 43, 63, 191n22, 202n20, 203n21, 204n33; causality and, 33, 45; Müller and, 33; nonconscious world and, 24; world of representation and, 31
Schrift, Alan, 63, 155
Schüttpelz, Erhard, 103
Schwann, Theodor, 28
science, xiii, 8, 23, 46, 69, 85, 111; capitalism and, 120; critique of, 49; deconstruction of, 47; gay, 50; sacralization of, 206; technology and, 57
science and technology studies (STS), 22
scientific discoveries, xiii, 8, 111
Searle, John, 138
second law of thermodynamics, 29, 190
self, xx, 125, 133, 135, 165; autonomous, 139; becoming of, 129; modern, 127, 130; speech and, 136; subjectivity and, 146
self-consciousness, 89, 131, 214n26
self-organization, biological/social/material, 72
self-preservation, 157, 215n33
self-reflection, 10–11
self-styling, 127–31, 189n46
sensibility, 117, 118, 122; hyperhumanist, 169
sensory apparatus, 37, 58
sensory perception (*Empfindung*), 27, 29, 32, 36, 192n24
Sharon, Tamar, 7, 21–22, 51, 188n43, 188n44, 221n63; biopolitics and, 181; cartography of, 24, 215n29; justification framework and, 181; posthumanism and, 23, 53, 125, 188n42
Shrift, Alan, 63, 155, 213n11
Siegert, Bernhard, 103, 208n12
Siemens, Herman, 139, 151, 155, 213n11, 215n33
Simonin, David, 199n32
Singularity Is Near, The (Kurzweil), 212n56
situatedness, 129, 130, 135, 172, 178, 182

slavery, 69, 73, 74, 80, 85, 198n27, 203n29
Sleigh, Charlotte, 62, 196n11, 196n12, 199n31
Sloterdijk, Peter, 4
Smith, Adam, 161
social aggregates, xviii, 85
social change, 9, 56, 98, 175
social contentiousness, 128–29
social dynamics, 76, 149, 172
social emergence, 79, 94
social existence, 20, 42, 153
social formations, 72, 76, 85, 165, 169, 175, 176; evolution of, 170–71
social interactions, xxii, 160, 168, 206n53
sociality, xxii, 70, 79, 96, 159, 180; animal, 76; conceptualizing, 76
social order, 66, 92, 154, 203n29
social organization, 174, 175
social phenomena, 9, 10, 94
social practices, 129, 141
social processes, 64, 76, 137, 169, 178, 180
social relations, 99, 145, 154, 161
social structures, 70, 76, 93, 172
social systems, xxii, 160, 176–77
social theories, 9, 98
social units, 71, 143
Sociological Enlightenment (Luhmann), 15
sociology, xxiii, xix, 59, 60, 71, 178
Socrates, xv, 16
Soper, Kate, xi
Sorgner, Stefan, 52, 123, 186n8
soul, understanding of, 213n15
sovereign individual, xix, 110, 124, 132, 134, 138, 139, 149, 210n35, 215n34, 216n35; creation of, 127, 131; herd and, xx
specific nerve energies, principle of, 27–30
speculative realism, 7, 24, 26, 53, 55, 195n60
Spencer, Herbert, 62, 63, 71, 169, 198–99n29, 202n20; Darwin/social realm and, 170; influence of, 70; social theory of, 198n29
Spencer, Richard B., 152
spiders, 63, 64, 67, 178, 185n3, 216n38; analogy, 197n21
Spinoza, Baruch, 147, 205n41
Stalinism, xv, 15, 156
Stegmaier, Werner, 179, 195n47, 205n49, 218n23, 220n42
Stiegler, Bernard, xix, 102, 109
stigmergy, 172, 220n43
Stone, Allucquère Rosanne, 188n44
subject formations, 173, 177
subjectivity, xx, 2, 135, 136, 137, 165, 195n64, 216n46; concept of, 189n46; human, 113, 125; notions of, 133, 189n46; self and, 146; understanding of, 213n15
"Supernormal Animal, The" (Massumi), 83
swarm, 72–77, 160; dynamics of, xii, 8, 58, 174; human, 165–76; model of, 169, 175, 176
systems theory, 7, 24, 50, 178, 194n46, 220n42

tactile game, 31, 32, 34
Taylor, Frederick, 202n20
technê, 105; *epistêmê* and, 208–9n17

technological determinism, 124
technological systems, xxii, 137, 160, 176–77, 178, 180
technology, 8, 9, 12, 99, 101, 103, 118, 120, 146, 154, 159, 163, 169, 180; advances in, 22, 111; affirmation of, 148; civilization and, 147; communication-media, xiii, xix, xx, 121, 172–73, 212n56; conception of, 98, 102; cultivation, 103, 141, 165; culture and, 111; definitions of, 122–23, 212n56; discipline as, 140–41; engagement with, 207n6; evolution of, 103, 123; history of, 105–11; hominization effects of, xiv, 6, 124, 127, 141–42, 149; humans and, 102, 109, 123, 206n3; impact of, xiii, 8, 207n6; language as, 131–37; media, 10, 98, 114, 118, 121; medical, 8; modern, 116, 164; nature and, 147, 148; philosophy of, xii–xiii, 207n6; posthumanism and, xx, 99, 110, 146–50, 216n40; question of, 101; reflections on, xii–xiii, xix, 112; role of, 110, 212n56; science and, xiii, 57; social practices and, 141; symbolic, 146; theorizing, 121; understanding of, 103, 110, 124, 130; writing, 111–14
"Technology, Environment and Social Risk" (Luhmann), 164
teleology, 75, 122, 123, 170
Thacker, Eugene, 61
Theraulaz, Guy, 60, 172
thermodynamics, xiii, 19, 49, 112
Thévenot, Laurent, 181
thinking, 56, 83; conscious, 60; dualist, 53, 94; human, 60, 145, 183; humanist, 19, 20; posthumanist, 83; unconscious, 82
Thompson, Evan, 30, 35, 51, 190n14
Thomsen, Mads Rosendahl, 11
Thoreau, Henry David, 205n41
Thousand Plateaus, A (Deleuze and Guattari), 76
Thus Spoke Zarathustra (Nietzsche), 85, 110, 145, 155, 174, 210n35, 212n56
togetherness, xxii, 159, 168, 174, 182
transhumanism, 2, 5, 7, 9, 14, 21, 22, 52, 53, 122, 195n57, 206n2, 212n55, 212n56; nature/technology and, 148; Nietzsche and, 186n8; overhuman and, 3; strands of, 123, 216n40
"Transhumanist Manifesto" (Vita-More), 14
triumph of the will, 95, 96
Trump, Donald, 152
truth, 43, 69; concept of, 42, 66, 154; concerns, 154; intellect/thing and, 42; metaphors and, 68; power and, 25
Twilight of the Idols (Nietzsche), 194n41, 205n45
typewriter, writing technologies and, 111–14

Über die phantastischen Gesichterscheinungen/On Imagined Visual Representations (Müller), 28
Übermensch, xi, 139, 218n23; cyborgs and, 121–24
Uexküll, Jakob von, 190n8
ultimate human (*der letzte Mensch*), 110

ultimate symbol (*Letztsymbol*), 51
unconscious apprehension (*unbewusste Anschauung*), 89
unconscious judgments (*unbewusste Schlüsse*), 89
universality, 12, 96
Unthought (Hayles), 54
Untimely Meditation (Nietzsche), 96–97, 110, 112, 114, 211n43, 211n45, 215–16n35
Ure, Michael, 161, 219n30

Varela, Francisco J., xvi, 190n14; enactive/radical embodiment and, 30, 35, 51
Verbeek, Peter-Paul, 22
Vibrant Matter (Bennett), xviii
violence, xv, xxi, 144, 149, 162
Virchow, Rudolf, 27, 28
Vischer, Friedrich Theodor, 192n27
Visman, Cornelia, 103
vitalism, 158–59; material, xviii, 44, 205n41; prophetic, 177
Vita-More, Natasha, 14
Vogt, Johannes Gustav, 87
Volkelt, Johannes, 192n27
von Bülo, Bernhard, 75
von Gersdorff, Carl, 43
von Nägle, Carl von, 70
Vorlesungen über die Menschen- und Thierseele/Lectures on the human and animal soul (Wundt), 87–88, 202n18

Wagner, Richard, 202n20
Waite, Geoff, 207n6, 220n40
Wamberg, Jacob, 11
Wanderer and His Shadow, The (Nietzsche), 65, 101, 151, 198n24
Warren, Mark, 217n9
Weatherby, Leif, 108, 195n60, 209n22
Weimar Classicism, 16, 19
Wellbery, David, 49, 111, 221n50
Welzer, Harald, 221n56
Werber, Niels, 61
What Is Posthumanism? (Wolfe), 3, 53, 61
Wheeler, William Morton, 62, 197n12
"When Digital Health Meets Digital Capitalism" (Sharon), 181
Why We Have Never Been Posthuman (Latour), 12
will, 79, 91, 98–99; concept of, 32, 80, 85–94; deconstruction of, xviii, 94; disgregation of, xvii–xix, 205n45; flashes of, 97; freedom of, 90, 97; herd and, 94–98; memory of, 138; unconscious, 204n32; understanding of, 55, 64; willpower and, 95
willpower, 77, 203n32; attribution of, 90; efficacy of, 165; will and, 95
Wills, David, 208n12
will to power, 94, 171, 195n64; concept of, 80, 93, 213n13
Wilson, Edward O., 62, 197n12
Winckelmann, Johann Joachim, 16
Winthrop-Young, Geoffrey, 103, 208n12
Wolfe, Cary, xv, 4, 7, 22, 53, 61, 101, 158, 208n12, 215n29, 216n40; on biopolitics, 167; cultural studies and, 208n11; Esposito

and, 159; Hayles and, 195n57; posthumanism and, 3, 12, 18, 56, 146, 215n30; on subjectivity, 136; technology and, 104

Wolff, Christian, 88

Woodward, William, 30

work ethic, 75, 200n46

working class: exploitation of, 120; impossibility of, xix; inner value of, 73

World Health Organization, 208n15

Wundt, Wilhelm, xviii, 28, 54, 63, 202n20, 203n22, 204n37; contentions of, 88; monistic perspective and, 93; psychology/physiology and, 93, 94; Schneider and, 87–88; will and, 55, 85–94

Zarathustra, 174, 175, 186n8, 216n38

Žižek, Slavoj, 10

Zur vergleichenden Physiologie des Gesichtssinnes des Menschen und der Thiere/On the comparative study of the physiology of vision in humans and animals (Müller), 28, 37

Zylinska, Joanna, 188n44

CARY WOLFE, SERIES EDITOR
(continued from p. ii)

56 *Thinking Plant Animal Human: Encounters with Communities of Difference*
David Wood

55 *The Elements of Foucault*
Gregg Lambert

54 *Postcinematic Vision: The Coevolution of Moving-Image Media and the Spectator*
Roger F. Cook

53 *Bleak Joys: Aesthetics of Ecology and Impossibility*
Matthew Fuller and Olga Goriunova

52 *Variations on Media Thinking*
Siegfried Zielinski

51 *Aesthesis and Perceptronium: On the Entanglement of Sensation, Cognition, and Matter*
Alexander Wilson

50 *Anthropocene Poetics: Deep Time, Sacrifice Zones, and Extinction*
David Farrier

49 *Metaphysical Experiments: Physics and the Invention of the Universe*
Bjørn Ekeberg

48 *Dialogues on the Human Ape*
Laurent Dubreuil and Sue Savage-Rumbaugh

47 *Elements of a Philosophy of Technology: On the Evolutionary History of Culture*
Ernst Kapp

46 *Biology in the Grid: Graphic Design and the Envisioning of Life*
Phillip Thurtle

45 *Neurotechnology and the End of Finitude*
Michael Haworth

44 *Life: A Modern Invention*
Davide Tarizzo

43 *Bioaesthetics: Making Sense of Life in Science and the Arts*
Carsten Strathausen

42 *Creaturely Love: How Desire Makes Us More and Less Than Human*
Dominic Pettman

41 *Matters of Care: Speculative Ethics in More Than Human Worlds*
María Puig de la Bellacasa

40 *Of Sheep, Oranges, and Yeast: A Multispecies Impression*
Julian Yates

39 *Fuel: A Speculative Dictionary*
Karen Pinkus

38 *What Would Animals Say If We Asked the Right Questions?*
Vinciane Despret

37 *Manifestly Haraway*
Donna J. Haraway

36 *Neofinalism*
Raymond Ruyer

35 *Inanimation: Theories of Inorganic Life*
David Wills

34 *All Thoughts Are Equal: Laruelle and Nonhuman Philosophy*
John Ó Maoilearca

33 *Necromedia*
Marcel O'Gorman

32 *The Intellective Space: Thinking beyond Cognition*
Laurent Dubreuil

31 *Laruelle: Against the Digital*
Alexander R. Galloway

30 *The Universe of Things: On Speculative Realism*
Steven Shaviro

29 *Neocybernetics and Narrative*
Bruce Clarke

28 *Cinders*
Jacques Derrida

27 *Hyperobjects: Philosophy and Ecology after the End of the World*
Timothy Morton

26 *Humanesis: Sound and Technological Posthumanism*
David Cecchetto

25 *Artist Animal*
Steve Baker

24 *Without Offending Humans: A Critique of Animal Rights*
Élisabeth de Fontenay

23 *Vampyroteuthis Infernalis: A Treatise, with a Report by the Institut Scientifique de Recherche Paranaturaliste*
Vilém Flusser and Louis Bec

22 *Body Drift: Butler, Hayles, Haraway*
Arthur Kroker

21 *HumAnimal: Race, Law, Language*
Kalpana Rahita Seshadri

20 *Alien Phenomenology, or What It's Like to Be a Thing*
Ian Bogost

19 *CIFERAE: A Bestiary in Five Fingers*
Tom Tyler

18 *Improper Life: Technology and Biopolitics from Heidegger to Agamben*
Timothy C. Campbell

17 *Surface Encounters: Thinking with Animals and Art*
Ron Broglio

16 *Against Ecological Sovereignty: Ethics, Biopolitics, and Saving the Natural World*
Mick Smith

15 *Animal Stories: Narrating across Species Lines*
Susan McHugh

14 *Human Error: Species-Being and Media Machines*
Dominic Pettman

13 *Junkware*
    Thierry Bardini

12 *A Foray into the Worlds of Animals and Humans,* with *A Theory of Meaning*
    Jakob von Uexküll

11 *Insect Media: An Archaeology of Animals and Technology*
    Jussi Parikka

10 *Cosmopolitics II*
    Isabelle Stengers

 9 *Cosmopolitics I*
    Isabelle Stengers

 8 *What Is Posthumanism?*
    Cary Wolfe

 7 *Political Affect: Connecting the Social and the Somatic*
    John Protevi

 6 *Animal Capital: Rendering Life in Biopolitical Times*
    Nicole Shukin

 5 *Dorsality: Thinking Back through Technology and Politics*
    David Wills

 4 *Bíos: Biopolitics and Philosophy*
    Roberto Esposito

 3 *When Species Meet*
    Donna J. Haraway

 2 *The Poetics of DNA*
    Judith Roof

 1 *The Parasite*
    Michel Serres

Edgar Landgraf is distinguished research professor of German at Bowling Green State University in Ohio. He is author of *Improvisation as Art: Conceptual Challenges, Historical Perspectives* and coeditor of *Posthumanism in the Age of Humanism: Mind, Matter, and the Life Sciences after Kant* and *Play in the Age of Goethe: Theories, Narratives, and Practices of Play around 1800*.